# Advances in Animal Science and Zoology

*Advances in Animal Science and Zoology* publishes original research on a wide range of topics that include everything from the production, management, and care of livestock and domestic animals to the biology, evolution, and conservation of all animal life and natural ecologies. Papers on all aspects of the animal kingdom will be considered for publication in Advances in Animal Science and Zoology.

*Advances in Animal Science and Zoology* aims to publish no less than one volume per year. The book series is owned and operated by Nova Science Publishers and is funded entirely by subscription revenue and author-optional publication services. The volumes are available in print and electronic formats.

nova
science publishers

www.novapublishers.com

# Advances in Animal Science and Zoology

**Advances in Animal Science and Zoology. Volume 20**
Owen P. Jenkins (Editor)
2022. ISBN: 979-8-88697-199-6 (Hardcover)
2022. ISBN: 979-8-88697-217-7 (eBook)

**Advances in Animal Science and Zoology. Volume 19**
Owen P. Jenkins (Editor)
2022. ISBN: 978-1-68507-522-4 (Hardcover)
2022. ISBN: 978-1-68507-534-7 (eBook)

**Advances in Animal Science and Zoology. Volume 18**
Owen P. Jenkins (Editor)
2021. ISBN: 978-1-68507-255-1 (Hardcover)
2021. ISBN: 978-1-68507-298-8 (eBook)

**Advances in Animal Science and Zoology. Volume 17**
Owen P. Jenkins (Editor)
2021. ISBN: 978-1-53619-838-6 (Hardcover)
2021. ISBN: 978-1-53619-862-1 (eBook)

**Advances in Animal Science and Zoology. Volume 16**
Owen P. Jenkins (Editor)
2020. ISBN: 978-1-53618-713-7 (Hardcover)
2020. ISBN: 978-1-53618-727-4 (eBook)

**Advances in Animal Science and Zoology. Volume 15**
Owen P. Jenkins (Editor)
2020. ISBN: 978-1-53618-254-5 (Hardcover)
2020. ISBN: 978-1-53618-283-5 (eBook)

More information about this series can be found at
https://novapublishers.com/product-category/series/advances-in-animal-science-and-zoology/

**Owen P. Jenkins**
Editor

# Advances in Animal Science and Zoology

**Volume 21**

nova science publishers

www.novapublishers.com

## NOTICE TO THE READER

The Publisher has taken reasonable care in the preparation of this book but makes no expressed or implied warranty of any kind and assumes no responsibility for any errors or omissions. No liability is assumed for incidental or consequential damages in connection with or arising out of information contained in this book. The Publisher shall not be liable for any special, consequential, or exemplary damages resulting, in whole or in part, from the readers' use of, or reliance upon, this material. Any parts of this book based on government reports are so indicated and copyright is claimed for those parts to the extent applicable to compilations of such works.

Independent verification should be sought for any data, advice or recommendations contained in this book. In addition, no responsibility is assumed by the Publisher for any injury and/or damage to persons or property arising from any methods, products, instructions, ideas or otherwise contained in this publication.

This publication is designed to provide accurate and authoritative information with regards to the subject matter covered herein. It is sold with the clear understanding that the Publisher is not engaged in rendering legal or any other professional services. If legal or any other expert assistance is required, the services of a competent person should be sought. FROM A DECLARATION OF PARTICIPANTS JOINTLY ADOPTED BY A COMMITTEE OF THE AMERICAN BAR ASSOCIATION AND A COMMITTEE OF PUBLISHERS.

**Library of Congress Cataloging-in-Publication Data**

ISBN: 979-8-89113-291-7

Published by Nova Science Publishers, Inc. † New York

# Contents


| | | |
|---|---|---|
| **Preface** | | **vii** |
| **Chapter 1** | **Canine Primary Intracranial Tumors: A Review**<br>Tomás Rodrigues Magalhães and Felisbina Luísa Queiroga | **1** |
| **Chapter 2** | **Balancing Vitamin A Supply for Cattle: A Review of the Current Knowledge**<br>Yauheni Shastak and Wolf Pelletier | **33** |
| **Chapter 3** | **The Influence of Geochemical Characteristics and Geological Formations on Mammal Density and Distribution: A Case Study from Gorny Altai, Russia**<br>Alexander V. Shitov, Andrey V. Karanin and Rachel A. Grant | **55** |
| **Chapter 4** | **Nutritional Influences on the Population Dynamics of Adjacent Rocky Mountain Bighorn Sheep Populations Showing Differing Trends**<br>Octavio C. Rosas-Rosas and Louis C. Bender | **77** |
| **Chapter 5** | **Macrocyclic Lactone-Based Tri-Model Therapy to Treat Humpsore in Cattle under a Humid Tropical Island Ecosystem**<br>P. Perumal A. K. De and D. Bhattacharya | **97** |
| **Chapter 6** | **Is the Decrease in cAMP Concentration in Oocytes Really a Trigger for Oocyte Maturation Induction? - From the Perspective of Fish Research**<br>Md. Forhad Hossain and Toshinobu Tokumoto | **117** |

**Chapter 7** **Looking Backwards and Forwards to the Rights of All Animals: Analogical Natural Law and 'Total Liberation' ........................................ 139**
Michael Allen

**Chapter 8** **The Effect of Urbanization on Carabidae and Scarabaeidae Families in Eskişehir Osmangazi University's Meşelik Campus and Meşelik Forest Gradient ........................................ 153**
Ebru Ceren Fidan, D. Ümit Şirin and Hilal Deniz Çelik

**Chapter 9** **The Morphological Characteristics of Female *Haemonchus Contortus* Obtained from Goats in the State of Maranhão, Brazil ........................................ 175**
Danilo Rodrigues Barros Brito, Joanice Francisca Machado Araújo, Livio Martins Costa Júnior and Rafael Arturo Torres-Fajardo

**Commentary** **Don't Liberate the Animals; Let them Resist! ............ 187**
Michael Allen

**Contents of Earlier Volumes ........................................ 195**

**Index ........................................ 201**

# Preface

This volume contains ten chapters that detail advances in animal science and zoology. Chapter One focuses on primary intracranial tumors in dogs, highlighting the three most prevalent types: meningiomas, gliomas, and choroid plexus tumors. Chapter Two aims to review the complex subject of vitamin A supplies for cattle. Chapter Three reveals significant correlations between underlying geology and the abundance of some species of mammal, even when other factors such as soil type, climate and vegetation are held constant. Chapter Four compares the diets and diet quality of two adjacent Rocky Mountain bighorn sheep herds with dissimilar population trends in Rocky Mountain National Park, Colorado. Chapter Five discusses macrocyclic lactone-based tri-model therapy used to treat humpsore in cattle. Chapter Six summarizes research on the involvement of changes in intracellular cAMP concentrations in the fish and frog oocyte maturation-inducing pathway. Chapter Seven discusses natural law as a neglected resource in the discourse of animal rights. Chapter Eight studies the effect of urbanization on the Carabidae and Scarabaeidae families in Eskişehir Osmangazi University's Meşelik Campus and Meşelik Forest gradient. Chapter Nine presents the morphological characteristics of female Haemonchus contortus. Lastly, Chapter Ten discusses animal liberation versus animal resistance.

# Chapter 1

# Canine Primary Intracranial Tumors: A Review

**Tomás Rodrigues Magalhães[1,2]**
**and Felisbina Luísa Queiroga[1,2,3,*]**
[1]Department of Veterinary Sciences, University of Trás-os-Montes and Alto Douro, Vila Real, Portugal
[2]Animal and Veterinary Research Centre (CECAV), University of Trás-os-Montes and Alto Douro, Vila Real, Portugal
[3]Center for the Study of Animal Sciences, CECA-ICETA, University of Porto, Porto, Portugal

## Abstract

This chapter will focus on primary intracranial tumors in the dog, highlighting three most prevalent types: meningiomas, gliomas, and choroid plexus tumors. These nervous system tumors originate from intracranial structures such as the meninges and brain parenchyma and can be classified as extra-axial or intra-axial, respectively. The neuroanatomical and histopathological classification of these tumors will be addressed in the chapter. Most intracranial tumors occur in middle-aged to older dogs, with a greater representation of retrievers (Labrador and Golden), Boxers, Boston terriers, French bulldogs and German shepherds, and it is assumed that there is no sexual predisposition overall. The most common clinical manifestation is seizures and although the definitive diagnosis is reached only through histopathology, most tumors are presumptively diagnosed through imaging (computed tomography or magnetic resonance imaging). Although complete surgical excision of the tumor is considered the ideal therapeutic approach, in most cases it is

* Corresponding Author's Email: fqueirog@utad.pt.

In: Advances in Animal Science and Zoology. Volume 21
Editor: Owen P. Jenkins
ISBN: 979-8-89113-291-7

not possible due to the neuroanatomical location of the tumor and its invasive properties. Thus, radiotherapy has been considered the treatment of choice as a single modality or in association with other therapies, such as surgery or chemotherapy, providing longer survival times than other treatment options. Furthermore, in order to identify new approaches to tumor control and improve the prognosis of these patients, novel therapies have been investigated, such as gene and hormone therapy and immunotherapy. Therefore, throughout the chapter, the clinical presentation and the diagnostic approach of these three types of canine intracranial tumors will be addressed, and their various treatment options and respective prognoses will be also discussed at the end.

**Keywords**: choroid plexus tumor, dog, glioma, literature review, meningioma

## Introduction

The incidence of canine brain tumors was described in the 1980s already as approximately 14.5 cases per 100 000 patients (Vandevelde, 1984). However, no clear or objective etiology has yet been described (Morris & Dobson, 2001).

Meningioma represents 45%-50% of primary intracranial tumors in dogs, being the most frequent tumor type in this group (Snyder et al., 2006; Staudinger et al., 2022). This neoplasm of meningeal origin develops outside the brain parenchyma (extra-axial) and is associated, in most of its variants, with a benign behavior (Motta et al., 2012; Horta et al., 2015). At clinical presentation, patients with meningioma have a median age of 10 years, with cases reported from 1 to 16 years (Montoliu et al., 2006; Snyder et al., 2006; Sturges et al., 2008). It should be noted that, compared to other intracranial tumors, patients with this tumor type are usually significantly older (Snyder et al., 2006; Schwartz et al., 2011). Breed predisposition has been described for meningioma with dolichocephalic breeds, especially retrievers (Labrador and Golden) and German shepherds, being the most represented (Snyder et al., 2006; Sturges et al., 2008; Suñol et al., 2017). In addition, the Boxer is also overrepresented, according to several published studies (Snyder et al., 2006; Sturges et al., 2008). There is apparently no sexual predisposition for this type of tumors (Montoliu et al., 2006; Snyder et al., 2006), although there is a study that raised the hypothesis that meningiomas have a more aggressive behavior in females (Théon et al., 2000).

In turn, the second most common type of primary canine brain neoplasm is glioma, with a relative representation of 31%, which is differentiated into 2 main subtypes, astrocytoma (17%) and oligodendroglioma (14%), the two variants addressed in this chapter (Snyder et al., 2006). Gliomas develop in the brain parenchyma (intra-axial) and are associated with shorter survival times compared to meningiomas (Hu et al., 2015; Suñol et al., 2017; Schwarz et al., 2018), although a recent study showed a comparable survival outcome for both tumor types after radiotherapy (Magalhães et al., 2021). At presentation, the median age of these patients is 8 years, ranging between 1.5 and 15 years (Snyder et al., 2006; Young et al., 2011; Koehler et al., 2018; Moirano et al., 2018; José-López et al., 2021; Hidalgo Crespo et al., 2022). Like meningiomas, gliomas also have a breed predisposition, being associated, in this case, with brachycephalic breeds, particularly Boxers and Boston terriers, but no sexual predisposition was identified in most studies (Snyder et al., 2006; Stoica et al., 2011; Young et al., 2011; Bentley et al., 2018; Moirano et al., 2018; José-López et al., 2021; Hidalgo Crespo et al., 2022). However, Koehler et al. (2018) in an article reviewing the diagnostic classification of canine glioma, found a male predilection even when neutered dogs were considered. This interesting finding needs to be investigated in future studies to ascertain this potential sexual influence.

Finally, choroid plexus tumor represents 7%-10% of primary brain tumors in dogs and develops within the ventricular cavities, particularly inside the fourth ventricle, from the choroid plexus epithelium (Snyder et al., 2006; Nentwig et al., 2012; Dalton et al., 2019; Staudinger et al., 2022). Previous studies described a median age of 5 years for choroid plexus papilloma (CPP) and 7 years for choroid plexus carcinoma (CPC), with cases reported from 3 to 14 years, and no sexual predisposition (Westworth et al., 2008; Nentwig et al., 2012; Reginato et al., 2016). Large breeds appear to be predisposed, with one study describing an overrepresentation of Golden retrievers (Westworth et al., 2008).

## Central Nervous System: Anatomical and Functional Features

Before approaching, in more detail, the classification of intracranial tumors and for a better understanding of their neuroanatomical location and respective potential brain damage, we will now review the anatomical and functional features of the central nervous system (CNS).

The nervous system can be divided into two main components: a central component, which includes the brain and spinal cord, and a peripheral component, which includes peripheral nerves and ganglia (König et al., 2008).

Regarding the CNS, the brain is the most important anatomical location for this chapter.

Five different brain areas can be distinguished, such as the telencephalon, diencephalon, mesencephalon (or midbrain), metencephalon and myelencephalon (Figure 1). Sometimes these areas appear grouped according to their primitive development. Thus, the first two constitute the forebrain and the latter, the hindbrain (König et al., 2008).

The telencephalon, commonly known as the brain, is considered the largest region of the CNS, consisting of two cerebral hemispheres separated by the longitudinal fissure. Symmetrically distributed in these two hemispheres, there are four pairs of main lobes, with distinct neurological functions: the frontal, the temporal, the occipital and the parietal lobes (Evans & de Lahunta, 2010). In a simplified way, the frontal lobe is responsible for voluntary motor activities and learning, the temporal lobe for hearing, the occipital lobe for vision and the parietal lobe for proprioception and nociception. Other structures are also part of the telencephalon, such as the hippocampus, the piriform lobes, the olfactory bulbs and peduncles and, subcortically, the basal nuclei (Lorenz & Kornegay, 2004). The diencephalon is organized into four zones: epithalamus, thalamus, metathalamus and hypothalamus. The region of the epithalamus, where the pineal gland is located, is responsible for regulating the circadian rhythm and events associated with sexual maturity. In turn, the thalamus functions as the reception and integration center for sensory information, which will later be transmitted to the cortical level, and the metathalamus, through its geniculate bodies, is responsible for receiving the optic and auditory pathways.

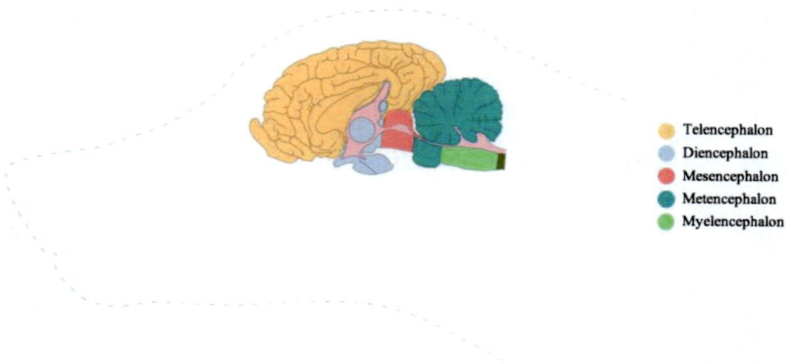


**Figure 1.** Representative diagram of the five canine brain areas.

Finally, the hypothalamus is the center of the autonomic nervous system and also plays an important role in neuroendocrine coordination, through the functional axis that connects it to the pituitary gland (König et al., 2008). The diencephalon is also associated with olfactory and visual functions through cranial nerves I and II, respectively, and with emotional behaviors due to its connections to the limbic system (Garosi, 2004).

The mesencephalon comprises the tectum, the cerebral peduncles, and the mesencephalic aqueduct and tegmentum. These structures are interconnected as follows: the mesencephalic aqueduct connects the ventricles ($3^{rd}$ and $4^{th}$), and the tegmentum connects the tectum with the cerebral peduncles. The functional component of this brain segment is associated with cranial nerves III and IV, whose nuclei are located here. Thus, at the midbrain level, auditory and visual reflexes are produced, in association with the colliculi, as well as the motor coordination, controlled by the red nucleus present in this region (König et al., 2008).

The metencephalon is composed by the pons and the cerebellum (König et al., 2008). The first is responsible for the motor activity of the masticatory muscles, as well as the sensory innervation of the face and jaw, through the cranial nerve V (Lorenz & Kornegay, 2004). In turn, the cerebellum structurally consists of two hemispheres and a protuberance in the middle called the vermis and is functionally responsible for balance, motor coordination, muscle tone regulation and also has an inhibitory action on urination (Garosi, 2004; König et al., 2008).

Finally, in the myelencephalon, there is the medulla oblongata, which is responsible for regulating basic organic functions, such as breathing and circulation, but also reflex activity, since it is in this location that the nuclei of cranial nerves VI to XII, involved in a variety of sensory and motor functions, are present (König et al., 2008).

At the cellular level, nervous tissue is composed by neurons, as the basic and functional unit of the nervous system, and glial cells or neuroglia, responsible for their structural and functional maintenance. Different cell types are part of the neuroglia, such as astrocytes, oligodendrocytes, ependymal cells and microglia, which perform various functions in association with the neuronal segment, such as communication and support, control of the extracellular environment (astrocytes), electrical isolation through the myelin production (oligodendrocytes), promotion of cerebrospinal fluid flow (ciliated ependymal cells) and, finally, phagocytic activity (microglia) (Junqueira & Carneiro, 2008).

Covering the brain, as well as the spinal cord, are three connective tissue membranes, called the meninges. The outermost one, called the dura mater or pachymeninge, is responsible for the interface between the bone surface of the skull and the other membranes, through its union with the periosteum (König et al., 2008). Below the dura mater is the arachnoid, followed by the pia mater, which communicates with nerve cells through astrocyte extensions (Junqueira & Carneiro, 2008). The last two are called leptomeninges and the space between them is called the subarachnoid space where the cerebrospinal fluid (CSF) circulates. CSF provides structural and metabolic support as it protects the CNS from mechanical aggressions and allows the nutrition and the elimination of products resulting from neuronal cell metabolism (König et al., 2008). This fluid is produced in the choroid plexus that line the ventricles and is absorbed by the villi present on the surface of the arachnoid membrane in an uninterrupted process (Junqueira & Carneiro, 2008).

## Classification of Canine Nervous System Tumors

Koestner et al. (1999) proposed a classification system for tumors of the nervous system of domestic animals, according to histological differentiation (Table 1). Tumors were divided into eight main groups: tumors of neuroepithelial tissue, tumors of the meninges, lymphomas and hematopoietic tumors, tumors of the sellar region, other primary tumors and cysts, metastatic tumors, local extensions of regional tumors and tumors of the peripheral nervous system. For the purpose of this chapter, only the first two groups will be addressed, as gliomas and choroid plexus tumors are part of the first group and meningiomas of the second one (Koestner et al., 1999).

Intracranial tumors can also be defined as primary or secondary depending on whether they originate from nervous tissue or other organic tissues, either by extension from neighboring regions (e.g., nasal cavity) or by distant metastasis through the bloodstream from any other location (Morris & Dobson, 2001). The most prevalent secondary tumor described is hemangiosarcoma, but pituitary tumors, lymphosarcoma and metastatic carcinomas are also frequent (Snyder et al., 2008). Most of these lesions tend to appear in the brain.

**Table 1.** Histological classification of tumors of the nervous system of domestic animals (Koestner et al., 1999)

| | |
|---|---|
| 1. Tumors of neuroepithelial tissue | 1.1. Astrocytic tumors<br>Low-grade (well differentiated), medium-grade (anaplastic) or high-grade (glioblastoma) astrocytomas<br>1.2. Oligodendroglial tumors<br>Oligodendroglioma or anaplastic (malignant) oligodendroglioma<br>1.3. Other gliomas<br>Mixed glioma (oligoastrocytoma), gliosarcoma, gliomatosis cerebri or spongioblastoma<br>1.4. Ependymal tumors<br>Ependymoma or anaplastic (malignant) ependymoma<br>1.5. Choroid plexus tumors<br>Choroid plexus papilloma or choroid plexus carcinoma<br>1.6. Neuronal and mixed neuronal-glial tumors<br>Gangliocytoma, ganglioglioma or olfactory neuroblastoma (esthesioneuroblastoma)<br>1.7. Embryonal tumors<br>Primitive neuroectodermal tumors (pnet), neuroblastoma, ependymoblastoma or thoracolumbar spinal cord tumor of young dogs<br>1.8. Pineal parenchymal tumors<br>Pineocytoma or pineoblastoma |
| 2. Tumors of the meninges | 2.1. Tumors of the meningothelial cells<br>Meningioma<br>2.2. Mesenchymal, nonmeningothelial tumors<br>Fibrosarcoma or diffuse meningeal sarcomatosis |
| 3. Lymphomas and hematopoietic tumors | 3.1. Lymphoma (lymphosarcoma)<br>3.2. Non-b, non-t leukocytic neoplasm (neoplastic reticulosis)<br>3.3. Microgliomatosis<br>3.4. Malignant histiocytosis |
| 4. Tumors of the sellar region | 4.1. Suprasellar germ cell tumor<br>4.2. Pituitary adenoma<br>4.3. Pituitary carcinoma<br>4.4. Craniopharyngioma |
| 5. Other primary tumors and cysts | 5.1. Vascular hamartoma<br>5.2. Epidermoid cyst<br>5.3. Pituitary cyst<br>5.4. Other cysts |
| 6. Metastatic tumors | |
| 7. Local extensions of regional tumors | 7.1. Nasal carcinoma<br>7.2. Multilobular tumor of bone<br>7.3. Chordoma |
| 8. Tumors of the peripheral nervous system | 8.1. Ganglioneuroma<br>8.2. Peripheral neuroblastoma<br>8.3. Paraganglioma<br>8.4. Peripheral nerve sheath tumor<br>Benign (schwannoma, neurofibroma) or malignant (malignant schwannoma, neurofibrosarcoma) |

One of the explanations for the current higher incidence of secondary tumors in the intracranial region is the greater "temporal opportunity" for their development, as recent advances in cancer therapy allow considerably higher survival rates for these animals (Snyder et al., 2008).

## Pathophysiology and Clinical Signs

The pathogenesis of these tumors is associated with their expansion and growth process, which leads to the destruction of the nervous tissue where they are located, as well as damage caused by continuous compression on adjacent tissues. Secondary to these processes, other events can be observed, such as cerebral edema, increased intracranial pressure and even cerebral herniation, which reflect tumor progression and lead to several clinical signs. Therefore, the clinical course of intracranial neoplastic disease usually proceeds slowly and progressively over time (LeCouteur & Withrow, 2007). Regarding the metastatic ability of brain tumors, the most described is the intraventricular or subarachnoid dissemination of neoplastic cells from choroid plexus tumors called drop metastases (Westworth et al., 2008; Reginato et al., 2016). On the other side, the appearance of metastases outside the intracranial cavity is rare (Coates & Johnson, 2010; Mariani, 2011), but still possible, as secondary pulmonary involvement has already been described in the literature (Schulman et al., 1992). One of the reasons given for this lower tendency of distant neoplastic distribution is the reduced exfoliation capacity of these tumors, due to their ability to adhere to adjacent tissues (Schulman et al., 1992).

There are several clinical signs that result from this type of intracranial lesions, such as, in order of greater to lesser frequency described, seizures, changes in mental status, vestibular signs, blindness, neck pain, anisocoria, tremors of the head and/or limbs and circling (Snyder et al., 2006). These clinical signs often only become evident after the tumor has reached a considerable size (Dennis, 2003). Seizure is the most frequently reported clinical sign in patients with intracranial tumors, particularly in dogs diagnosed with meningioma and glioma, although there is no significant correlation between the type of tumor and the development of this neurological manifestation (Snyder et al., 2006; Schwartz et al., 2011; Mariani et al., 2015). Seizures can be generalized or focal, with the first condition being the most frequent (Schwartz et al., 2011). These animals tend to have recurrent episodes, establishing a chronic neurological disease, defined as secondary or symptomatic epilepsy (Podell, 2004). Possible factors that explain the onset of epileptic activity secondary to the tumor process are cerebral ischemia, ionic changes, denervation of cerebral cortical regions and neurotransmitter imbalance, for example. Epilepsy management is based on the administration of antiepileptic drugs, which can be combined with corticosteroids to reduce peritumoral edema and the occurrence risk of these manifestations (Schwartz et al., 2011). Imepitoin, levetiracetam, phenobarbital, potassium bromide, felbamate and gabapentin are examples of these drugs, with the first two being associated with greater safety due to the fewer reported adverse effects resulting from their administration (Charalambous et al., 2016). Still, it should be noted that these clinical signs, secondary to brain tumors, can be difficult to control, as it is often not possible to eliminate the underlying mechanisms and causes (Schwartz et al., 2011).

All clinical signs are dependent on the specific tumor location (Morris & Dobson, 2001). For example, seizures are related to the cerebral cortex, particularly the frontal lobe, while cranial nerve disorders and paresis are associated with brain stem, dysmetria and ataxia to the cerebellum, changes in mentation to the hypothalamus and, finally, head tilt, circling and nystagmus to the central vestibular area (Morris & Dobson, 2001; Schwartz et al., 2011).

## Diagnosis: Presumptive versus Definitive

From the history and clinical signs and even before starting the diagnostic exploration, it is possible to define the differential diagnoses. Thus, it is reasonable to assume that the various types of brain neoplasms are differential diagnoses of each other, but there are still other non-tumor diseases that could be considered as such due to identical clinical presentations and similar disease courses. Cerebrovascular diseases, infectious diseases of bacterial, viral, fungal or parasitic origin, inflammatory diseases, such as granulomatous meningoencephalitis, and metabolic and toxic brain conditions are some examples (Troxel, 2015). Moreover, as seizure is the most common clinical presentation of these tumors, idiopathic epilepsy will also have to be ruled out (North & Banks, 2009; Mariani, 2011). To rule out these possible pathological conditions and reach a final diagnosis, it is necessary to use complementary exams, particularly imaging.

## Computed Tomography and Magnetic Resonance Imaging

Computed tomography (CT) and magnetic resonance imaging (MRI) are two imaging tools that allow the establishment of a presumptive diagnosis of brain tumors. The diagnostic value of the two modalities is different: CT is preferred for the evaluation of bone tissue abnormalities, while MRI is better for the detection of soft tissue lesions (Dennis, 2003). Specifically, for brain tumors, MRI has a higher diagnostic value and is therefore the most recommended approach (Kraft et al., 1997; LeCouteur, 1999; Dennis, 2003). Despite its qualities, MRI does not always allow a diagnosis and can even lead to inaccuracies (Ródenas et al., 2011; Wolff et al., 2012). For example, Ródenas et al. (2011) described an accuracy of 70% for the diagnosis of primary brain tumors through MRI, showing that this technique has some limitations that can lead to incorrect determination of the tumor type. Moreover, Leclerc et al., (2013) conducted a study evaluating the interobserver agreement regarding several imaging features in MRI scans performed on canine patients. Good results were obtained for findings such as the mass effect and the presence and intensity of enhancement, but also for the establishment of an accurate diagnosis. However, some common features of intracranial neoplasms, such as the pattern of enhancement, the cerebral edema and the dural tail sign, were determined with higher variability, resulting in a lower degree of agreement. Despite this, MRI remains the imaging method of choice for the presumptive diagnosis of brain tumors (Ródenas et al., 2011).

In their most typical form, meningiomas correspond to well-differentiated and broad-based masses with a spherical to ovoid shape and with smooth to irregular margins (Kraft et al., 1997). Most meningiomas are hypo- or isointense on T1-weighted images (T1WI) and iso- or hyperintense on T2-weighted images (T2WI) (Thomas et al., 1996; Kraft et al., 1997; Snyder et al., 2006; Sturges et al., 2008). Regarding the signal uniformity, it is particularly heterogeneous in T1WI and homogeneous in T2WI. The presence of edema is frequent (about 94% of cases) and its appearance is variable as it can be located around the tumor or in a more diffuse pattern (Sturges et al., 2008). Contrast enhancement is uniformly strong (Kraft et al., 1997; Wisner et al., 2011) and its pattern is often heterogeneous (Thomas et al., 1996; Snyder et al., 2006). The most common location of these tumors is the telencephalon (Ródenas et al., 2011) and, within this, the olfactory region is the most frequently described (Snyder et al., 2006; Sturges et al., 2008) along with the frontal lobes (Mariani, 2011). Mass effect is described as a structural midline shift associated or not with ventricular compression (Ródenas et al., 2011) and is usually present in a subtle form in these neoplasms (Thomas et al., 1996). Although not pathognomonic, the dural tail sign is closely associated with meningiomas (Graham et al., 1998; Sturges et al., 2008; Wisner et al., 2011) and can be identified on T1WI as a "linear enhancement of thickened dura mater adjacent to an extra-axial mass", after contrast administration (Graham et al., 1998).

In turn, gliomas are intra-axial and well-defined masses, with a round to ovoid shape. Most of these neoplasms are uniformly hypointense on T1WI and hyperintense on T2WI compared to gray matter (Thomas et al., 1996; Snyder et al., 2006; Young et al., 2011; José-López et al., 2021). In these tumors, the presence of edema is variable, but is usually defined as minimal to moderate, whereas the mass effect is often moderate to severe (Thomas et al., 1996). Contrast enhancement is most often associated with high-grade gliomas and its pattern is generally heterogeneous or even ring-like (Thomas et al., 1996; Snyder et al., 2006; Wisner et al., 2011; Young et al., 2011). One difference noted between the two main types of glioma is their contact with the brain surface, which is more frequent in oligodendrogliomas, probably due to a greater ability for peripheral invasion (Young et al., 2011). Regarding location, gliomas tend to appear in the prosencephalon (Stoica et al., 2011; Young et al., 2011), and the temporal and piriform lobes are two regions where this type of tumor is frequently detected (Mariani, 2011).

Choroid plexus tumors also have particular imaging features on MRI, presenting as intraventricular masses with a frequent globular shape and associated with ventriculomegaly (Westworth et al., 2008). Most of these tumors are iso- or hyperintense and associated with a marked and homogenous contrast enhancement on T1WI, and hyperintense on T2WI (Kraft et al., 1997; Westworth et al., 2008; Antonakakis et al., 2022). Periventricular edema is another reported finding that has been described more frequently in CPCs than in CPPs (69% vs 44%), which may help in the imaging differentiation of these two tumor subtypes (Westworth et al., 2008).

On MRI, other common signs can be found in these three tumor types, which allows their characterization and the reinforcement of the respective diagnosis. Cystic areas, hemorrhage, mineralization, necrosis and hyperostosis are some examples reported in the literature (Ródenas et al., 2011; Motta et al., 2012). Cystic lesions correspond to well-defined structures composed of fluid accumulation or necrotic areas (Cherubini et al., 2005) and were described in 26% of the meningiomas analyzed in a previous study (Sturges et al., 2008). Although Kraft et al. (1997) have not identified any association between the presence of cystic lesions and a specific tumor type, more recent studies have described that these and necrotic regions are more frequently observed in glial neoplasms (Ródenas et al., 2011; Young et al., 2011). In turn, hyperostosis results from bone erosion caused by the presence of the tumor in a process called signal void (Motta et al., 2012). Another important aspect is the possibility of characterizing the tumor in terms of its behavior, based on the changes observed on MRI. For example, irregular tumor margins, tumor invasion and hemorrhage are signs associated with neoplasms with a more aggressive and malignant behavior (Kraft et al., 1997).

The imaging features of these neoplasms on CT are quite similar to those reported for MRI. On CT, meningiomas are identified as isodense or hyperdense lesions, in relation to the remaining brain tissue, whereas gliomas can also be hypodense. After contrast administration, enhancement is homogeneously strong in meningeal and choroid plexus tumors, while in glial tumors it can be more variable (absent to strong) and in a homogeneous, heterogeneous or ring-like pattern, such as in MRI. Pathological changes common to these tumor types, such as edema and mass effect, are also detected with this imaging method, as well as more specific findings such as dural tail sign and hyperostosis (Hecht, 2011).

## Histopathology, Immunochemistry and Biomarkers

Although most of these neoplasms are diagnosed using the imaging methodologies described above, in fact the only definitive diagnostic method is the histopathological examination of the tumor through lesion biopsy, usually at the time of surgery (North & Banks, 2009; Motta et al., 2012). Advances have been made to improve the biopsy technique in these intracranial locations, and imaging (CT or MRI) has been used for this purpose. In this way, the stereotactic brain biopsy emerged in order to provide greater safety and precision to this procedure (Stoica et al., 2011; Dickinson, 2014).

Histologically, canine meningiomas can be differentiated into nine subtypes, according to the classification system proposed by the World Health Organization (WHO). Meningotheliomatous, fibrous, transitional, psammomatous, angiomatous, papillary, granular cell, myxoid and anaplastic meningiomas are the nine established histological variables, the first eight being considered as neoplasms of benign behavior and the last one of malignant behavior (Koestner et al., 1999). Another subtype that has been more recently identified and described is microcystic meningioma (Sturges et al., 2008). The meningothelial and transitional subtypes are the most frequently reported in the literature (Barnhart et al., 2002; Montoliu et al., 2006; Sturges et al., 2008). In human medicine, meningeal tumors are divided into more variants and are differentiated by malignancy grade into grade I or benign, grade II or atypical and grade III or anaplastic (Louis et al., 2016). Although this differentiation in histological grades is not defined in veterinary medicine, some authors use this distinction in comparison with human medicine. Therefore, Sturges et al. (2008) differentiated the meningiomas diagnosed in their study by grade, including in grade I, five previously mentioned meningioma variants: meningothelial, transitional, psamomatous, angiomatous and microcystic. Of the 112 meningiomas, 56% were grade I, 43% grade II and only 1% grade III, showing a higher prevalence of benign neoplasms. Meningiomas can also be evaluated by immunohistochemical staining using antibodies against vimentin, S100 protein, neuron-specific enolase (NSE), cytokeratin (CK) and glial fibrillary acidic protein (GFAP) (Montoliu et al., 2006). Of all of them, vimentin is noteworthy, as its immunolabeling was observed in all meningiomas tested in previous studies (Barnhart et al., 2002; Montoliu et al., 2006). Thus, immunohistochemistry allows confirmation of the diagnosis through the association between certain markers with this tumor type and even between the detected immunolabeling and the histological subtype (Barnhart et al., 2002; Montoliu et al., 2006).

As previously mentioned, it is possible to distinguish several glioma subtypes through histological features. For example, astrocytomas can be differentiated into low (well differentiated), medium (anaplastic), and high-grade tumors. Low-grade astrocytomas are defined as fibrillary, protoplasmic or gemistocytic, and high-grade astrocytomas are generally identified as glioblastoma (Koestner et al., 1999). Likewise, oligodendrogliomas can also be histopathologically differentiated into low grade or high grade (Koehler et al., 2018). Immunohistochemical staining mentioned above for meningiomas is also used for gliomas, and GFAP is particularly detected in astrocytic tumors (Stoica et al., 2011; Koehler et al., 2018; José-López et al., 2021). Positive Olig2 immunoreactivity has also been associated with canine gliomas (Koehler et al., 2018).

Although the histological tumor classification proposed by Koestner et al. (1999) distinguishes canine choroid plexus tumors into only two subtypes (CPP and CPC), an intermediate grade named as atypical choroid plexus papilloma (or grade II) is currently considered, according to the WHO classification system for human CNS neoplasms (Louis et al., 2016). For this differentiation, the mitotic index, the cellularity and the papillary pattern are the most significant histological features (Westworth et al., 2008; Reginato et al., 2016). Kir7.1 is the immunohistochemical marker with the most diagnostic value for these tumors, as it is highly specific for the choroid plexus epithelium, promoting an immunolabeling of the apical membrane of these neoplastic cells and allowing the differentiation of choroid plexus tumors from other primary or metastatic brain tumors (e.g., carcinomas that have metastasized to the CNS) (Choi et al., 2016). A positive and intense immunoreactivity for Kir7.1 was found in all choroid plexus tumors analyzed in previous studies (Choi et al., 2016; Dalton et al., 2019). GFAP, E-cadherin, N-cadherin and β-catenin immunoexpression has also been evaluated for tumor differentiation and diagnostic purposes, but with lower specificity and overall consistency (Nentwig et al., 2012; Hirose et al., 2015; Reginato et al., 2016).

Another tool used in tumor diagnosis are biomarkers, such as the vascular endothelial growth factor (VEGF) and its receptor (VEGFR), the epidermal growth factor receptor (EGFR) and the platelet-derived growth factor receptor (PDGFR), whose expression is increased in this type of neoplasms. It should be noted that VEGF and VEGFR are associated with carcinogenesis, vascular permeability change and tumor neovascularization and that they can be found in these three types of canine intracranial tumors (Dickinson, Roberts, et al., 2006; Stoica et al., 2011; Muscatello et al., 2018). A higher expression of these markers is detected in gliomas compared to meningiomas and within the latter, it is higher in those with higher tumor grade (Dickinson, Roberts, et al., 2006). Among gliomas, oligodendroglioma and, mainly, glioblastoma multiforme present greater expression. Thus, an association between a higher expression of these biomarkers and a higher degree of tumor malignancy has been considered in meningiomas and gliomas (Dickinson, Roberts, et al., 2006; Rossmeisl et al., 2007). In relation to choroid plexus tumors, this correlation between the expression of these factors and tumor grade was not identified (Muscatello et al., 2018). Moreover, this labeling detection may not only have a diagnostic value, but can also be important for the therapeutic protocol, since the use of therapies directed to these biological molecules has been hypothesized for the treatment of canine intracranial tumors (Dickinson, Roberts, et al., 2006; Rossmeisl et al., 2007; Stoica et al., 2011).

## Cerebrospinal Fluid Analysis

CSF analysis is another complementary diagnostic tool recommended in patients with brain disorders. This exam may reinforce the suspicion of neoplasia by ruling out inflammatory conditions (LeCouteur & Withrow, 2007). Increased total cell count and total protein concentration in CSF have been associated with the diagnosis of these brain tumors (Dickinson, Sturges, et al., 2006; Snyder et al., 2006; Ródenas et al., 2011). For example, gliomas have been associated with higher CSF cell counts compared to other tumor types (Ródenas et al., 2011; Moirano et al., 2018) and choroid plexus tumors with increased protein concentration (Westworth et al., 2008). In addition, Westworth et al. (2008) reported a significantly higher CSF protein concentration in CPCs compared to CPPs, with values greater than 80 mg/dL being associated only with the malignant variant, which may help in tumor differentiation. Two other CSF findings observed are mixed cell pleocytosis and albuminocytologic dissociation, with the first being considered the most common cytologic abnormality in canine patients with primary intracranial tumors (Snyder et al., 2006; Westworth et al., 2008; Moirano et al., 2018; José-López et al., 2021). Although much less common, neoplastic cells can also be found in CSF (Snyder et al., 2006; Westworth et al., 2008). Despite all these abnormalities, in a small percentage of cases (about 10% to 16%), the CSF analysis can be completely normal (Dickinson, Sturges, et al., 2006; Snyder et al., 2006). CSF collection should be performed carefully, since it will promote pressure changes in animals already prone to high intracranial pressure (ICP), which may predispose to brain herniation. In order to prevent this problem, before proceeding to CSF collection, a mannitol solution can be administered, and the patient must be hyperventilated in order to reduce the ICP (LeCouteur & Withrow, 2007).

## Other Diagnostic Tests

In addition to the main methods mentioned above, routine tests such as blood count, serum biochemistry and urinalysis should also be performed, and their results are usually normal for these intracranial neoplastic diseases (Troxel, 2015). Other complementary tests that can provide information about the primary tumor or the possibility of a different etiology for the neurological signs are thoracic radiography and abdominal ultrasound (LeCouteur & Withrow, 2007). These exams also allow the detection of other concomitant diseases, which may be decisive when establishing a therapeutic plan for the primary intracranial condition (Bigio Marcello et al., 2015). Skull radiographs are also an option, but their usefulness is limited only to the identification of tumors located in the nasal cavity or in the skull that also affect the brain region by local extension. Structural changes associated with these tumors, such as lysis, hyperostosis and mineralization can also be detected through this radiographic approach (LeCouteur & Withrow, 2007). Brainstem auditory evoked potential testing and electroencephalography are also described in the literature in order to localize the lesion in the auditory pathways or in the cerebrum, respectively (LeCouteur, 1999).

## Treatment

The treatment of canine intracranial tumors depends on the therapeutic goal, which can be defined as the palliative control of clinical signs only or the definitive neoplastic treatment to achieve a cure or at least increase patients' disease-free and survival times. This therapeutic decision depends on several factors such as the clinical presentation, the patient's characteristics, the presence of comorbidities, the owner's financial constraints and the type of tumor diagnosed, as well as its individual features and neuroanatomical location (Mariani, 2011).

Therefore, several therapeutic modalities can be chosen for single or combined management, such as symptomatic/ medical treatment, surgery, radiotherapy and chemotherapy. In addition, novel therapies have been investigated as also described at the end of this section.

### Symptomatic Treatment

Symptomatic treatment is based on the combination of several drugs, particularly anticonvulsants to control tumor-related epilepsy and corticosteroids to reduce peritumoral edema. Given the frequent need to manage these effects, this form of therapy is almost always present in combination with other therapeutic strategies (North & Banks, 2009). However, when more invasive therapies, such as surgery and radiotherapy, are not feasible due to different factors (e.g., individual tumor features or owner's financial constraints), this may be the only approach performed to control clinical signs (Motta et al., 2012).

Despite its proven effects and rare exceptions of better therapeutic success described in the literature (Itoh et al., 2016), it is known that patients only symptomatically treated have a poor prognosis, with a median survival time of 26 to 89 days, according to some studies from the last decade (Rossmeisl et al., 2013; Van Meervenne et al., 2014; Treggiari et al., 2017; Moirano et al., 2018; José-López et al., 2021).

## Surgery

Surgery can have several applications in the approach to canine brain tumors: as a curative technique, allowing the total tumor removal, as a way of reducing the tumor volume (debulking) and/or as an auxiliary diagnostic method by allowing the neoplastic mass biopsy (Troxel, 2015). In most cases, only the last two applications are carried out, since it is usually not possible to completely remove the tumor due to the invasive properties of this type of neoplasms (Evans et al., 1993). In fact, sometimes surgical reintervention is necessary to achieve a better outcome (Hidalgo Crespo et al., 2022). Patients with extra-axial neoplasms (such as meningioma) and whose tumor is located on the dorsal and/or lateral surfaces of the brain are considered better candidates for surgical intervention. However, even under these conditions, tumor recurrence is common when this treatment method is used alone due to the potential growth of the residual tumor population (Mariani, 2011). For example, meningiomas tend to recur up to 1-year post-craniotomy, if not associated with any other adjunctive therapy (Fossum, 2013). Animals with tumors located in the frontal lobes of the brain are an exception as they occasionally allow for a complete surgical recession (LeCouteur, 1999; Suñol et al., 2017).

Due to these limitations of traditional intracranial surgery, new techniques have been developed using specific equipment in order to allow a wider tumor excision and ensuring safer margins. The surgical aspirator (Greco et al., 2006) and the endoscope for intracranial use (Klopp & Rao, 2009) are two examples. The use of these new technologies represents an advantage in the survival time of these patients in comparison with animals submitted to traditional surgery or traditional surgery plus adjuvant radiotherapy (Greco et al., 2006; Klopp & Rao, 2009). Klopp and Rao (2009) reported that the endoscopic-assisted intracranial tumor removal technique, in addition to the survival advantage, also significantly reduces the animal's hospitalization time compared to animals that underwent post-surgical radiotherapy, resulting in a lower economic impact to the pet owner.

Surgical intervention for tumor recession can be performed through several approaches, particularly transfrontal (Greco et al., 2006; Klopp & Rao, 2009; Uriarte et al., 2011; Fossum, 2013; Suñol et al., 2017; Hidalgo Crespo et al., 2022), rostrotentorial (Greco et al., 2006; Klopp & Rao, 2009; Fossum, 2013) and suboccipital (Klopp & Rao, 2009; Fossum, 2013; Hidalgo Crespo et al., 2022). Others are also described, such as sinotemporal (Uriarte et al., 2011), transnasal or transcribriform, frontoparietal, parietotemporal, transpetrosal, and even through ventral brainstem and transoptic foramen (Klopp & Rao, 2009; Suñol et al., 2017). The main problem with surgical procedures is the possibility of leaving lesions in the surrounding normal tissues (Jeffery & Brearley, 1993), in addition to the risk of local bacterial infection, pneumocranium (Klopp & Rao, 2009), hemorrhage and increased ICP (Mariani, 2011). Nevertheless, some studies consider the morbidity and mortality associated with craniotomy to be "acceptably low" (Niebauer et al., 1991).

Hu et al., (2015) described in their systematic review a median survival time of 312 days for surgically treated canine intracranial tumors, with a range of 27 to 2104 days. It should be noted that for canine patients with meningioma, there is a greater variability in terms of survival depending on the surgical approach performed. For example, in cases where more advanced techniques were applied, such as the use of a surgical aspirator or an endoscope, the median survival times obtained were 41.2 months (3.4 years) and 70 months (5.8 years), respectively (Greco et al., 2006; Klopp & Rao, 2009). These survival times contrast significantly with those obtained using conventional surgical techniques, such as the mean value of 6.6 months (Niebauer et al., 1991) and the median of 7 months (Axlund et al., 2002).

A more recent study evaluated the survival time of dogs with histologically confirmed intracranial tumors treated with surgical resection alone (Suñol et al., 2017). The authors identified a statistically longer median survival time for patients with meningiomas compared with the glioma group (422 days versus 66 days), showing that meningiomas have a better response to the surgical approach. Likewise, the intraventricular location makes surgical excision of choroid plexus tumors difficult and challenging, so alternative approaches have been suggested recently in case reports with greater success, such as the transcallosal (Lehner et al., 2020) and the telovelar approaches (Antonakakis et al., 2022).

## Radiotherapy

There are several radiotherapy protocols recommended for the treatment of canine intracranial tumors with different treatment schedules, total doses and doses per fraction. The treatment schedule can be set to three days a week, such as Monday, Wednesday and Friday (M-W-F), during the five days of the week (M-F) or even weekly. Regarding total and fractional doses, these can range from 24 to 54 Gy and from 2 to 6 Gy, respectively (McEntee, 2004). A recommended and classic treatment protocol is to provide a total radiation dose of 45 Gy in 15 fractions over a 3-week period (LeCouteur & Withrow, 2007).

The radiotherapy protocols used can be divided into two types: definitive or palliative. While in the first type, the objective is to keep the animal in clinical remission or, if possible, the definitive resolution of the neoplastic lesion, in the second type, the purpose is only to minimize the clinical signs and improve the animal's quality of life (Pruitt & Thrall, 2011; Tollett et al., 2016). These two types of radiation administration regimens differ in total doses and per fraction, which influence the outcome, but also the expected potential adverse effects (McEntee, 2006).

There are several publications evaluating the survival times of dogs with brain tumors treated with radiotherapy. The median survival time described by Hu et al. (2015) in their systematic review was 351 days (≈ 11.5 months). Better results were obtained in more recently published studies, which shows that the advances in radiotherapy techniques and procedures have been improving the outcome of these patients, describing median survival times between 524-756 days (Treggiari et al., 2017; Schwarz et al., 2018; Van Asselt et al., 2020; Magalhães et al., 2021; Staudinger et al., 2022).

Therefore, radiotherapy is considered the most successful treatment method for these neoplastic conditions (Heidner et al., 1991; Evans et al., 1993; Hu et al., 2015) and can be associated with other therapies, such as pre- or post-surgical intervention (North & Banks, 2009). Axlund et al. (2002) described a significantly longer median survival time for dogs with intracranial meningioma treated with this combination compared to those who underwent surgical tumor recession alone at a ratio of 16.5 months to 7 months, respectively. A similar result was obtained in another study in dogs with meningiomas located in the frontal lobe, whose location is associated with a worse prognosis in the reviewed literature. In that study, Uriarte et al. (2011) showed that canine patients who underwent adjuvant radiotherapy after surgery had a median survival time of 18 months. Finally, Keyerleber et al.

(2015) reported that the median survival time for patients who underwent three-dimensional conformal radiotherapy (3D-CRT) alone and for patients treated with the combination with surgery was similar, highlighting the significant therapeutic effect of 3D-CRT and even suggesting it as a good alternative when surgery is not feasible.

## Chemotherapy

Chemotherapy has limited therapeutic application due to several features of intracranial tumors and their location. For example, the heterogeneity of the intra-tumoral cell population results in only a small part of the tumor being sensitive to the cytotoxic drug, without a constant and uniform therapeutic response (LeCouteur & Withrow, 2007). This limitation is mainly observed in intra-axial tumors, such as gliomas, which present different cell types, as well as histological variations (Jeffery & Brearley, 1993). Furthermore, the use of therapeutic doses at the tumor site may not be possible without causing damage to the surrounding tissues, which are considered dose-limiting. Finally, another relevant aspect is the blood-brain barrier, which prevents the anticancer drug passing from the bloodstream to the brain tissue, denying the adequate exposure to tumors and the consequent expected effect of these drugs (LeCouteur & Withrow, 2007). In order to overcome this problem, a possible approach that has been described to increase permeability to these drugs in the peritumoral region is to break this barrier by administering a hyperosmotic solution, such as mannitol. However, this procedure does not seem to present satisfactory results in brain tissues with neoplastic lesions, contrary to what is seen in normal tissues, given their different biology (Groothuis et al., 1990).

Although most of the results published to date do not show an advantage in its use, some previous reports advocate the role of chemotherapy in tumor control. Hydroxyurea, lomustine and more recently chlorambucil and temozolomide are examples of cytostatic drugs used for this purpose (Jeffery & Brearley, 1993; Tamura et al., 2007; Bentley et al., 2018; Hidalgo Crespo et al., 2022).

Satisfactory results were obtained in a case of canine intracranial meningioma, in which the combination of hydroxyurea with corticosteroid therapy was used as a single treatment (Tamura et al., 2007). The authors described a decrease in tumor volume associated with an improvement in clinical signs, showing the benefits of this combination therapy that allowed the suppression of tumor growth by decreasing tumor size and reducing perilesional edema.

The clinical effect of using lomustine is also evident in the current literature. For example, the administration of this alkylating agent has shown therapeutic benefits when administered as single treatment or in combination with conventional symptomatic drugs in previous studies (Jeffery & Brearley, 1993; Jung et al., 2006; Van Meervenne et al., 2014; Moirano et al., 2018). Van Meervenne et al. (2014) described a median survival time of 93 days when lomustine is given in conjunction with symptomatic treatment in dogs diagnosed with both intra- and extra-axial brain tumors, although the impact of this combination was not considered significant on overall patient survival. An opposite result was obtained by Moirano et al. (2018) who found a survival advantage for dogs diagnosed with glioma treated with adjuvant lomustine compared to those treated only symptomatically: 138 days versus 35 days, respectively. Differences in tumor diagnosis (intra- and extra-axial brain tumors versus only gliomas) and diagnostic method (CT versus MRI) may have led to a treatment of intracranial tumors less responsive to chemotherapy or even misdiagnosis in the former, as suggested by the authors.

Regarding chlorambucil, there is a published study that described a well-tolerated protocol of metronomic pre- and post-surgical chemotherapy with that cytotoxic drug associated with surgical microsurgical tumor resection and monthly oral lomustine (Bentley et al., 2018). In that study, the presence of intra-tumoral chlorambucil was detected, proving its ability to alter and penetrate the blood-brain barrier, which resulted in a median overall survival of 257 days. The authors attributed these results to the potential immunomodulatory and antiangiogenic effects of metronomic chlorambucil.

Finally, temozolomide adjuvant chemotherapy in association with surgical debulking was described in canine patients diagnosed with intracranial gliomas, recording a median survival time of 240 days (Hidalgo Crespo et al., 2022)

Moreover, a meta-analysis published in 2007 showed the survival benefit of adjuvant chemotherapy in human patients diagnosed with choroid plexus tumors, leading the authors to ultimately advise the use of cytostatic drugs in cases of incomplete surgical excision of CPCs (Wrede et al., 2007). Although these findings raise the possibility that similar results could potentially be obtained in canine patients, there is not the same evidence in veterinary medicine to date. Further studies in dogs with a definitive diagnosis of choroid plexus tumor are needed to clarify this significant adjuvant effect.

In summary, chemotherapy appears to have therapeutic benefit, but with only limited clinical effect when applied as a single strategy (Dickinson, 2014). Further prospective studies with less variability (e.g., with definitive diagnosis of a specific tumor type and standardization of chemotherapeutic protocols) are needed to ascertain this true effect.

## Novel Treatments

In addition to the therapeutic approaches mentioned above, which are widely established in current clinical practice, there are other possible therapeutic tools under investigation that may prove to be advantageous in the future.

Gene therapy through the introduction of suppressor or modulator genes or even viral agents is an example, as it can modify neoplastic cell function and consequently increase their sensitivity to conventional therapies (LeCouteur, 1999; Dickinson, 2014).

The potential impact of hormonal therapies in the treatment of canine meningiomas has also been speculated, such as those targeting progesterone and its receptors. This hypothesis has been raised since several studies have shown a high expression of progesterone receptors in these tumors (Théon et al., 2000; Adamo et al., 2003). For example, Théon et al. (2000) described a progesterone receptor immunoreactivity in 14 of 20 (70%) incompletely resected intracranial meningiomas. However, despite this significant presence, the same is not true for estrogen receptors, whose detection is lower (Adamo et al., 2003) or even null (Théon et al., 2000). An interesting finding is that this detection can also provide prognostic information as an inverse correlation between progesterone receptor expression and tumor malignancy has been described in canine meningiomas (Adamo et al., 2003). Likewise, the detection of proliferating cell nuclear antigen (PCNA) has prognostic value as it is inversely correlated with the expression of these receptors, allowing the differentiation between more or less aggressive tumors and with greater or lesser therapeutic sensitivity (Théon et al., 2000).

Immunotherapy has also been investigated with regard to its applicability in the treatment of canine intracranial tumors following the investigations already conducted in human medicine (Dickinson, 2014). For example, Andersen et al., (2013) described a vaccine immunotherapy protocol, using autologous tumor cell lysate combined with synthetic toll-like receptor ligands, in dogs diagnosed with meningioma that had previously undergone craniotomy. In that study, the authors identified a mechanism of antibody-dependent cell-mediated cytotoxicity against allogeneic meningioma cells after vaccination, demonstrating its therapeutic benefits in these canine patients and encouraging further investigation.

## Prognosis

As described above, the therapeutic approach chosen has a clear impact on the prognosis of dogs diagnosed with intracranial tumors and consequently on their survival. However, there are several other prognostic factors already identified in the literature.

Regarding the clinical presentation, dogs older than 8 years (Moirano et al., 2018), with more severe neurological manifestations (Heidner et al., 1991; Jeffery & Brearley, 1993; LeCouteur & Withrow, 2007; Mariani, 2011; Schwarz et al., 2018; Van Asselt et al., 2020) and with shorter duration of clinical signs before diagnosis (Moirano et al., 2018) were associated with a worse outcome. The neuroanatomic location of the tumor is also assumed to be a negative prognostic factor, particularly if intra-axial tumors (Brearley et al., 1999; Hu et al., 2015), infratentorial tumors (Rossmeisl et al., 2013) or meningiomas located in the caudal brain/brain stem (Klopp & Rao, 2009) are considered in comparison with extra-axial tumors, supratentorial tumors or forebrain meningiomas, respectively.

According to the literature, a worse prognosis can be anticipated in patients with tumor involvement of multiple brain regions (Heidner et al., 1991; LeCouteur & Withrow, 2007); neoplastic masses with irregular and poorly defined margins, T2-weighted heterogenous signal and in association of drop metastases on MRI (José-López et al., 2021); and an increased white blood cell count in CSF (Heidner et al., 1991; LeCouteur & Withrow, 2007). Furthermore, when diagnosis is considered late (Mariani, 2011), rapid tumor progression (Heidner et al., 1991) or recurrence (Motta et al., 2012) are observed and when tumors are classified as secondary (Heidner et al., 1991), the expected outcome is also worse.

Finally, based on the type of intracranial tumor diagnosed, a prognosis can be established, as meningiomas have a more favorable prognosis compared to other intracranial tumors (Heidner et al., 1991; Klopp & Rao, 2009; Hu et al., 2015; Suñol et al., 2017; Schwarz et al., 2018). Within meningeal tumors, those with lower expression of progesterone receptors (Théon et al., 2000; Adamo et al., 2003), a high $PF_{PCNA}$ index (Théon et al., 2000), and those classified as anaplastic meningiomas (Greco et al., 2006) have the worst prognosis due to their typically more aggressive tumor behavior.

## Conclusion

Intracranial tumors should always be considered in middle-aged to older dogs with a sudden history of seizures, which should be subject to full diagnostic exploration. Although currently the diagnosis of intracranial tumors is mostly presumptive through MRI, we believe that with the new brain biopsy techniques being developed, the definitive diagnosis through histopathology will be increasingly possible to obtain ante-mortem, which consequently will improve the therapeutic protocol design.

Radiotherapy, alone or in conjunction with surgery, is considered the best treatment modality for these intracranial tumors and given the most recent studies that describe new protocols and administration techniques, it is expected that its positive therapeutic effect in the management of this neoplastic disease may increase further in the future.

Finally, several aspects related to the clinical presentation, the tumor type and its progression, the tumor imaging characteristics and the expression of neoplastic markers and receptors allow the establishment of an individual prognosis, which is crucial to choose the most suitable treatment and to manage the owner's expectations.

## Disclaimer

None.

## References

Adamo, P. F., Cantile, C., & Steinberg, H. (2003). Evaluation of progesterone and estrogen receptor expression in 15 meningiomas of dogs and cats. *Am J Vet Res*, *64*(10), 1310-1318.

Andersen, B. M., Pluhar, G. E., Seiler, C. E., Goulart, M. R., SantaCruz, K. S., Schutten, M. M., Meints, J. P., O'Sullivan, M. G., Bentley, R. T., Packer, R. A., Thomovsky, S. A., Chen, A. V., Faissler, D., Chen, W., Hunt, M. A., Olin, M. R., & Ohlfest, J. R. (2013). Vaccination for invasive canine meningioma induces in situ production of antibodies capable of antibody-dependent cell-mediated cytotoxicity. *Cancer Res, 73*(10), 2987-2997.

Antonakakis, M. G., Carletti, B. E., Anselmi, C., McGrath, S., & Minguez, J. J. (2022). Use of a telovelar approach for complete resection of a choroid plexus tumor in a dog. *Vet Surg, 51*(8), 1273-1279.

Axlund, T. W., McGlasson, M. L., & Smith, A. N. (2002). Surgery alone or in combination with radiation therapy for treatment of intracranial meningiomas in dogs: 31 cases (1989-2002). *J Am Vet Med Assoc, 221*(11), 1597-1600.

Barnhart, K. F., Wojcieszyn, J., & Storts, R. W. (2002). Immunohistochemical staining patterns of canine meningiomas and correlation with published immunophenotypes. *Vet Pathol, 39*(3), 311-321.

Bentley, R. T., Thomovsky, S. A., Miller, M. A., Knapp, D. W., & Cohen-Gadol, A. A. (2018). Canine (Pet Dog) Tumor Microsurgery and Intratumoral Concentration and Safety of Metronomic Chlorambucil for Spontaneous Glioma: A Phase I Clinical Trial. *World Neurosurg, 116*, e534-e542.

Bigio Marcello, A., Gieger, T. L., Jiménez, D. A., & Granger, L. A. (2015). Detection of comorbidities and synchronous primary tumours via thoracic radiography and abdominal ultrasonography and their influence on treatment outcome in dogs with soft tissue sarcomas, primary brain tumours and intranasal tumours. *Vet Comp Oncol, 13*(4), 433-442.

Brearley, M. J., Jeffery, N. D., Phillips, S. M., & Dennis, R. (1999). Hypofractionated radiation therapy of brain masses in dogs: a retrospective analysis of survival of 83 cases (1991-1996). *J Vet Intern Med, 13*(5), 408-412.

Charalambous, M., Shivapour, S. K., Brodbelt, D. C., & Volk, H. A. (2016). Antiepileptic drugs' tolerability and safety--a systematic review and meta-analysis of adverse effects in dogs. *BMC Vet Res, 12*, 79.

Cherubini, G. B., Mantis, P., Martinez, T. A., Lamb, C. R., & Cappello, R. (2005). Utility of magnetic resonance imaging for distinguishing neoplastic from non-neoplastic brain lesions in dogs and cats. *Vet Radiol Ultrasound, 46*(5), 384-387.

Choi, E. J., Sloma, E. A., & Miller, A. D. (2016). Kir7.1 immunoreactivity in canine choroid plexus tumors. *J Vet Diagn Invest, 28*(4), 464-468.

Coates, J. R., & Johnson, G. C. (2010). Nervous System Neoplasia. In C. J. Henry & M. L. Higginbotham (Eds.), *Cancer Management in Small Animal Practice* (pp. 186-194). W.B. Saunders.

Dalton, M. F., Stilwell, J. M., Krimer, P. M., Miller, A. D., & Rissi, D. R. (2019). Clinicopathologic Features, Diagnosis, and Characterization of the Immune Cell Population in Canine Choroid Plexus Tumors [Original Research]. *Frontiers in Veterinary Science, 6.*

Dennis, R. (2003). Advanced imaging: indications for CT and MRI in veterinary patients. *In Practice, 25*, 243-254.

Dickinson, P. J. (2014). Advances in diagnostic and treatment modalities for intracranial tumors. *J Vet Intern Med, 28*(4), 1165-1185.

Dickinson, P. J., Roberts, B. N., Higgins, R. J., Leutenegger, C. M., Bollen, A. W., Kass, P. H., & LeCouteur, R. A. (2006). Expression of receptor tyrosine kinases VEGFR-1

(FLT-1), VEGFR-2 (KDR), EGFR-1, PDGFRalpha and c-Met in canine primary brain tumours. *Vet Comp Oncol*, *4*(3), 132-140.

Dickinson, P. J., Sturges, B. K., Kass, P. H., & LeCouteur, R. A. (2006). Characteristics of cisternal cerebrospinal fluid associated with intracranial meningiomas in dogs: 56 cases (1985-2004). *J Am Vet Med Assoc*, *228*(4), 564-567.

Evans, H. E., & de Lahunta, A. (2010). Guide to the Dissection of the Dog. In H. E. Evans & A. de Lahunta (Eds.), *Guide to the Dissection of the Dog* (7th ed., pp. 262-288). Saunders.

Evans, S. M., Dayrell-Hart, B., Powlis, W., Christy, G., & VanWinkle, T. (1993). Radiation therapy of canine brain masses. *J Vet Intern Med*, *7*(4), 216-219.

Fossum, T. W. (2013). Surgery of the Brain. In T. W. Fossum (Ed.), *Small Animal Surgery* (4th ed., pp. 1438-1466). Elsevier.

Garosi, L. (2004). Lesion localization and differential diagnosis. In *BSAVA Manual of Canine and Feline Neurology* (3rd ed., pp. 24-34). British Small Animal Veterinary Association.

Graham, J. P., Newell, S. M., Voges, A. K., Roberts, G. D., & Harrison, J. M. (1998). The dural tail sign in the diagnosis of meningiomas. *Vet Radiol Ultrasound*, *39*(4), 297-302.

Greco, J. J., Aiken, S. A., Berg, J. M., Monette, S., & Bergman, P. J. (2006). Evaluation of intracranial meningioma resection with a surgical aspirator in dogs: 17 cases (1996-2004). *J Am Vet Med Assoc*, *229*(3), 394-400.

Groothuis, D. R., Warkne, P. C., Molnar, P., Lapin, G. D., & Mikhael, M. A. (1990). Effect of hyperosmotic blood-brain barrier disruption on transcapillary transport in canine brain tumors. *J Neurosurg*, *72*(3), 441-449.

Hecht, S. (2011). Brain. In T. Schwarz & J. Saunders (Eds.), *Veterinary Computed Tomography* (pp. 185-195). Wiley-Blackwell.

Heidner, G. L., Kornegay, J. N., Page, R. L., Dodge, R. K., & Thrall, D. E. (1991). Analysis of survival in a retrospective study of 86 dogs with brain tumors. *J Vet Intern Med*, *5*(4), 219-226.

Hidalgo Crespo, E., Farré Mariné, A., Pumarola i Battle, M., Borrego Massó, J. F., & Luján Feliu-Pascual, A. (2022). Survival Time after Surgical Debulking and Temozolomide Adjuvant Chemotherapy in Canine Intracranial Gliomas. *Veterinary Sciences*, *9*(8), 427.

Hirose, N., Uchida, K., Matsunaga, S., Chambers, J. K., & Nakayama, H. (2015). Expression of cell adhesion molecules in canine choroid plexus tumors. *J Vet Med Sci*, *77*(2), 255-259.

Horta, R., Martins, B., Martins, G., & Marcinowska, A. (2015). Advances in diagnosis and management of intracranial and spinal meningiomas in dogs and cats. *Revista Portuguesa de Ciências Veterinárias*, *110*, 172-180.

Hu, H., Barker, A., Harcourt-Brown, T., & Jeffery, N. (2015). Systematic Review of Brain Tumor Treatment in Dogs. *J Vet Intern Med*, *29*(6), 1456-1463.

Itoh, T., Uchida, K., Nishi, A., Shii, H., Nagayoshi, T., & Sakamoto, H. (2016). Choroid plexus papilloma in a dog surviving for 15 months after diagnosis with symptomatic therapy. *J Vet Med Sci*, *78*(1), 167-169.

Jeffery, N., & Brearley, M. J. (1993). Brain tumours in the dog: Treatment of 10 cases and review of recent literature. *Journal of Small Animal Practice*, *34*(8), 367-372.

José-López, R., Gutierrez-Quintana, R., de la Fuente, C., Manzanilla, E. G., Suñol, A., Pi Castro, D., Añor, S., Sánchez-Masian, D., Fernández-Flores, F., Ricci, E., Marioni-Henry, K., Mascort, J., Matiasek, L. A., Matiasek, K., Brennan, P. M., & Pumarola, M. (2021). Clinical features, diagnosis, and survival analysis of dogs with glioma. *J Vet Intern Med*, *35*(4), 1902-1917.

Jung, D. I., Kim, H. J., Park, C., Kim, J. W., Kang, B. T., Lim, C. Y., Park, E. H., Sur, J. H., Seo, M. H., Hahm, D. H., & Park, H. M. (2006). Long-term chemotherapy with lomustine of intracranial meningioma occurring in a miniature schnauzer. *J Vet Med Sci*, *68*(4), 383-386.

Junqueira, L. C., & Carneiro, J. (2008). Tecido Nervoso. In L. C. Junqueira & J. Carneiro (Eds.), *Histologia Básica* (11th ed., pp. 153-181). Guanabara Koogan.

Keyerleber, M. A., McEntee, M. C., Farrelly, J., Thompson, M. S., Scrivani, P. V., & Dewey, C. W. (2015). Three-dimensional conformal radiation therapy alone or in combination with surgery for treatment of canine intracranial meningiomas. *Vet Comp Oncol*, *13*(4), 385-397.

Klopp, L. S., & Rao, S. (2009). Endoscopic-assisted intracranial tumor removal in dogs and cats: long-term outcome of 39 cases. *J Vet Intern Med*, *23*(1), 108-115.

Koehler, J. W., Miller, A. D., Miller, C. R., Porter, B., Aldape, K., Beck, J., Brat, D., Cornax, I., Corps, K., Frank, C., Giannini, C., Horbinski, C., Huse, J. T., O'Sullivan, M. G., Rissi, D. R., Mark Simpson, R., Woolard, K., Shih, J. H., Mazcko, C., Gilbert, M. R., & LeBlanc, A. K. (2018). A Revised Diagnostic Classification of Canine Glioma: Towards Validation of the Canine Glioma Patient as a Naturally Occurring Preclinical Model for Human Glioma. *J Neuropathol Exp Neurol*, *77*(11), 1039-1054.

Koestner, A., Bilzer, T., Fatzer, R., F. Y., S., Summers, B. A., & Van Winkle, T. J. (1999). *Histological Classification of Tumors of the Nervous System of Domestic Animals* (2nd ed.). Armed Forces Institute of Pathology and American Registry of Pathology.

König, H. E., Liebich, H. G., & Cerveny, C. (2008). Sistema nervioso. In H. E. König & H. G. Liebich (Eds.), *Anatomía de los Animales Domésticos: Órganos, Sistema circulatorio y sistema nervioso* (pp. 203-276). Editorial Medica Panamericana.

Kraft, S. L., Gavin, P. R., DeHaan, C., Moore, M., Wendling, L. R., & Leathers, C. W. (1997). Retrospective review of 50 canine intracranial tumors evaluated by magnetic resonance imaging. *J Vet Intern Med*, *11*(4), 218-225.

Leclerc, M. K., d'Anjou, M. A., Blond, L., Carmel É, N., Dennis, R., Kraft, S. L., Matthews, A. R., & Parent, J. M. (2013). Interobserver agreement and diagnostic accuracy of brain magnetic resonance imaging in dogs. *J Am Vet Med Assoc*, *242*(12), 1688-1695.

LeCouteur, R. A. (1999). Current concepts in the diagnosis and treatment of brain tumours in dogs and cats. *J Small Anim Pract*, *40*(9), 411-416.

LeCouteur, R. A., & Withrow, S. J. (2007). Tumors of the Nervous System. In S. J. Withrow & D. M. Vail (Eds.), *Withrow & MacEwen's Small Animal Clinical Oncology (Fourth Edition)* (pp. 659-685). W.B. Saunders.

Lehner, L., Czeibert, K., Benczik, J., Jakab, C., & Nagy, G. (2020). Transcallosal Removal of a Choroid Plexus Tumor From the Lateral Ventricle in a Dog. Case Report. *Front Vet Sci*, *7*, 536.

Lorenz, M. D., & Kornegay, J. N. (2004). Localization of Lesions in the Nervous System. In M. D. Lorenz & J. N. Kornegay (Eds.), *Handbook of Veterinary Neurology* (4th ed., pp. 45-74). W.B. Saunders.

Louis, D. N., Perry, A., Reifenberger, G., von Deimling, A., Figarella-Branger, D., Cavenee, W. K., Ohgaki, H., Wiestler, O. D., Kleihues, P., & Ellison, D. W. (2016). The 2016 World Health Organization Classification of Tumors of the Central Nervous System: a summary. *Acta Neuropathol*, *131*(6), 803-820.

Magalhães, T. R., BenoÎt, J., Néčová, S., North, S., & Queiroga, F. L. (2021). Outcome After Radiation Therapy in Canine Intracranial Meningiomas or Gliomas. *In Vivo*, *35*(2), 1117-1123.

Mariani, C. (2011). Tumours of the nervous system. In J. M. Dobson & B. D. X. Lascelles (Eds.), *BSAVA Manual of Canine and Feline Oncology* (3rd ed., pp. 329-340). British Small Animal Veterinary Association.

Mariani, C. L., Schubert, T. A., House, R. A., Wong, M. A., Hopkins, A. L., Barnes Heller, H. L., Milner, R. J., Lester, N. V., Lurie, D. M., Rajon, D. A., Friedman, W. A., & Bova, F. J. (2015). Frameless stereotactic radiosurgery for the treatment of primary intracranial tumours in dogs. *Vet Comp Oncol*, *13*(4), 409-423.

McEntee, M. C. (2004). A survey of veterinary radiation facilities in the United States during 2001. *Vet Radiol Ultrasound*, *45*(5), 476-479.

McEntee, M. C. (2006). Veterinary radiation therapy: review and current state of the art. *J Am Anim Hosp Assoc*, *42*(2), 94-109.

Moirano, S. J., Dewey, C. W., Wright, K. Z., & Cohen, P. W. (2018). Survival times in dogs with presumptive intracranial gliomas treated with oral lomustine: A comparative retrospective study (2008-2017). *Veterinary and Comparative Oncology*, *16*(4), 459-466.

Montoliu, P., Añor, S., Vidal, E., & Pumarola, M. (2006). Histological and immunohistochemical study of 30 cases of canine meningioma. *J Comp Pathol*, *135*(4), 200-207.

Morris, J., & Dobson, J. (2001). Nervous System. In J. Morris & J. Dobson (Eds.), *Small Animal Oncology* (1st ed., pp. 192-203). Blackwell Science.

Motta, L., Mandara, M. T., & Skerritt, G. C. (2012). Canine and feline intracranial meningiomas: an updated review. *Vet J*, *192*(2), 153-165.

Muscatello, L. V., Avallone, G., Serra, F., Seuberlich, T., Mandara, M. T., Sisó, S., Brunetti, B., & Oevermann, A. (2018). Glomeruloid Microvascular Proliferation, Desmoplasia, and High Proliferative Index as Potential Indicators of High Grade Canine Choroid Plexus Tumors. *Vet Pathol*, *55*(3), 391-401.

Nentwig, A., Higgins, R. J., Francey, T., Doherr, M., Zurbriggen, A., & Oevermann, A. (2012). Aberrant E-cadherin, β-catenin, and glial fibrillary acidic protein (GFAP) expression in canine choroid plexus tumors. *J Vet Diagn Invest*, *24*(1), 14-22.

Niebauer, G. W., Dayrell-Hart, B. L., & Speciale, J. (1991). Evaluation of craniotomy in dogs and cats. *J Am Vet Med Assoc*, *198*(1), 89-95.

North, S., & Banks, T. (2009). Tumours of the Nervous System. In S. North & T. Banks (Eds.), *Introduction to Small Animal Oncology* (pp. 243-252). Saunders Elsevier.

Podell, M. (2004). Seizures. In S. R. Platt & N. J. Olby (Eds.), *BSAVA Manual of Canine and Feline Neurology* (3rd ed., pp. 97-112). British Small Animal Veterinary Association.

Pruitt, A. F., & Thrall, D. E. (2011). Principles of radiation therapy. In J. M. Dobson & B. D. X. Lascelles (Eds.), *BSAVA Manual of Canine and Feline Oncology* (3ª ed., pp. 80-90). British Small Animal Veterinary Association.

Reginato, A., Girolami, D., Menchetti, L., Foiani, G., & Mandara, M. T. (2016). E-cadherin, N-cadherin Expression and Histologic Characterization of Canine Choroid Plexus Tumors. *Vet Pathol*, *53*(4), 788-791.

Ródenas, S., Pumarola, M., Gaitero, L., Zamora, A., & Añor, S. (2011). Magnetic resonance imaging findings in 40 dogs with histologically confirmed intracranial tumours. *Vet J*, *187*(1), 85-91.

Rossmeisl, J. H., Duncan, R. B., Huckle, W. R., & Troy, G. C. (2007). Expression of vascular endothelial growth factor in tumors and plasma from dogs with primary intracranial neoplasms. *Am J Vet Res*, *68*(11), 1239-1245.

Rossmeisl, J. H., Jr., Jones, J. C., Zimmerman, K. L., & Robertson, J. L. (2013). Survival time following hospital discharge in dogs with palliatively treated primary brain tumors. *J Am Vet Med Assoc*, *242*(2), 193-198.

Schulman, F. Y., Ribas, J. L., Carpenter, J. L., Sisson, A. F., & LeCouteur, R. A. (1992). Intracranial meningioma with pulmonary metastasis in three dogs. *Vet Pathol*, *29*(3), 196-202.

Schwartz, M., Lamb, C. R., Brodbelt, D. C., & Volk, H. A. (2011). Canine intracranial neoplasia: clinical risk factors for development of epileptic seizures. *J Small Anim Pract*, *52*(12), 632-637.

Schwarz, P., Meier, V., Soukup, A., Drees, R., Besserer, J., Beckmann, K., Roos, M., & Rohrer Bley, C. (2018). Comparative evaluation of a novel, moderately hypofractionated radiation protocol in 56 dogs with symptomatic intracranial neoplasia. *J Vet Intern Med*, *32*(6), 2013-2020.

Snyder, J. M., Lipitz, L., Skorupski, K. A., Shofer, F. S., & Van Winkle, T. J. (2008). Secondary intracranial neoplasia in the dog: 177 cases (1986-2003). *J Vet Intern Med*, *22*(1), 172-177.

Snyder, J. M., Shofer, F. S., Van Winkle, T. J., & Massicotte, C. (2006). Canine intracranial primary neoplasia: 173 cases (1986-2003). *J Vet Intern Med*, *20*(3), 669-675.

Staudinger, C., Meier, V., Beckmann, K., Körner, M., & Rohrer Bley, C. (2022). Treatment of intracranial neoplasia in dogs using higher doses: A randomized controlled trial comparing a boosted to a conventional radiation protocol. *J Vet Intern Med*, *36*(4), 1353-1364.

Stoica, G., Levine, J., Wolff, J., & Murphy, K. (2011). Canine astrocytic tumors: a comparative review. *Vet Pathol*, *48*(1), 266-275.

Sturges, B. K., Dickinson, P. J., Bollen, A. W., Koblik, P. D., Kass, P. H., Kortz, G. D., Vernau, K. M., Knipe, M. F., Lecouteur, R. A., & Higgins, R. J. (2008). Magnetic resonance imaging and histological classification of intracranial meningiomas in 112 dogs. *J Vet Intern Med*, *22*(3), 586-595.

Suñol, A., Mascort, J., Font, C., Bastante, A. R., Pumarola, M., & Feliu-Pascual, A. L. (2017). Long-term follow-up of surgical resection alone for primary intracranial rostrotentorial tumors in dogs: 29 cases (2002-2013). *Open Vet J*, *7*(4), 375-383.

Tamura, S., Tamura, Y., Ohoka, A., Hasegawa, T., & Uchida, K. (2007). A canine case of skull base meningioma treated with hydroxyurea. *J Vet Med Sci*, *69*(12), 1313-1315.

Théon, A. P., Lecouteur, R. A., Carr, E. A., & Griffey, S. M. (2000). Influence of tumor cell proliferation and sex-hormone receptors on effectiveness of radiation therapy for dogs with incompletely resected meningiomas. *J Am Vet Med Assoc*, *216*(5), 701-707, 684-705.

Thomas, W. B., Wheeler, S. J., Kramer, R., & Kornegay, J. N. (1996). Magnetic resonance imaging features of primary brain tumors in dogs. *Veterinary Radiology & Ultrasound*, *37*(1), 20-27.

Tollett, M. A., Duda, L., Brown, D. C., & Krick, E. L. (2016). Palliative radiation therapy for solid tumors in dogs: 103 cases (2007-2011). *J Am Vet Med Assoc*, *248*(1), 72-82.

Treggiari, E., Maddox, T. W., Gonçalves, R., Benoit, J., Buchholz, J., & Blackwood, L. (2017). Retrospective comparison of three-dimensional conformal radiation therapy vs. prednisolone alone in 30 cases of canine infratentorial brain tumors. *Vet Radiol Ultrasound*, *58*(1), 106-116.

Troxel, M. T. (2015). Brain Neoplasia. In E. Côte (Ed.), *Clinical Veterinary Advisor: Dogs and Cats* (pp. 140-141). Elsevier.

Uriarte, A., Moissonnier, P., Thibaud, J. L., Reyes-Gomez, E., Devauchelle, P., & Blot, S. (2011). Surgical treatment and radiation therapy of frontal lobe meningiomas in 7 dogs. *Can Vet J*, *52*(7), 748-752.

Van Asselt, N., Christensen, N., Meier, V., Rohrer Bley, C., Laliberte, S., Poirier, V. J., Desai, N., Chen, Y., & Turek, M. (2020). Definitive-intent intensity-modulated radiation therapy provides similar outcomes to those previously published for definitive-intent three-dimensional conformal radiation therapy in dogs with primary brain tumors: A multi-institutional retrospective study. *Vet Radiol Ultrasound*, *61*(4), 481-489.

Van Meervenne, S., Verhoeven, P. S., de Vos, J., Gielen, I. M., Polis, I., & Van Ham, L. M. (2014). Comparison between symptomatic treatment and lomustine supplementation in 71 dogs with intracranial, space-occupying lesions. *Vet Comp Oncol*, *12*(1), 67-77.

Vandevelde, M. (1984, September 5-6). Brain tumors in domestic animals: an overview. *Proceedings of the Conference of Brain Tumors in Man and Animals,* North Carolina.

Westworth, D. R., Dickinson, P. J., Vernau, W., Johnson, E. G., Bollen, A. W., Kass, P. H., Sturges, B. K., Vernau, K. M., Lecouteur, R. A., & Higgins, R. J. (2008). Choroid plexus tumors in 56 dogs (1985-2007). *J Vet Intern Med*, *22*(5), 1157-1165.

Wisner, E. R., Dickinson, P. J., & Higgins, R. J. (2011). Magnetic resonance imaging features of canine intracranial neoplasia. *Vet Radiol Ultrasound*, *52*(1 Suppl 1), S52-61.

Wolff, C. A., Holmes, S. P., Young, B. D., Chen, A. V., Kent, M., Platt, S. R., Savage, M. Y., Schatzberg, S. J., Fosgate, G. T., & Levine, J. M. (2012). Magnetic resonance imaging for the differentiation of neoplastic, inflammatory, and cerebrovascular brain disease in dogs. *J Vet Intern Med*, *26*(3), 589-597.

Wrede, B., Liu, P., & Wolff, J. E. (2007). Chemotherapy improves the survival of patients with choroid plexus carcinoma: a meta-analysis of individual cases with choroid plexus tumors. *J Neurooncol*, *85*(3), 345-351.

Young, B. D., Levine, J. M., Porter, B. F., Chen-Allen, A. V., Rossmeisl, J. H., Platt, S. R., Kent, M., Fosgate, G. T., & Schatzberg, S. J. (2011). Magnetic resonance imaging features of intracranial astrocytomas and oligodendrogliomas in dogs. *Vet Radiol Ultrasound*, *52*(2), 132-141.

# Chapter 2

# Balancing Vitamin A Supply for Cattle: A Review of the Current Knowledge

**Yauheni Shastak*, PhD**
**and Wolf Pelletier, PhD**
BASF SE, Nutrition & Health, Ludwigshafen am Rhein, Germany

## Abstract

Among its many functions, vitamin A, also known as retinol, plays an essential role in growth, vision, immunity, cellular metabolism, bone development, and reproduction. Recent debate about retinol supplementation in commercial animal production is likely a result of its physiological importance, but also its high cost and susceptibility to oxidation. In cattle, inadequate vitamin A can reduce performance and increase susceptibility to diseases and reproductive problems. There is much to consider when determining an optimal vitamin A supply. Physiological condition, health, and nutritional status all affect retinol requirements in dairy and beef cattle. Estimated adequate vitamin A intake by scientific committees such as the National Research Council (NRC) or the National Academies of Sciences, Engineering, and Medicine (NASEM) consider the needs of healthy animals under optimal environmental conditions. In commercial production conditions, however, a higher vitamin supply is required for optimum weight gain, feed conversion, and health. Compounding balanced diets and achieving maximum performance in cattle requires an understanding of how vitamin A affects metabolism and health. This paper aims to review this complex subject, which has been rather neglected to date.

---

* Corresponding Author's Email: yauheni.shastak@basf.com.

In: Advances in Animal Science and Zoology. Volume 21
Editor: Owen P. Jenkins
ISBN: 979-8-89113-291-7

**Keywords:** vitamin A, retinol, cattle, supplementation, requirement

## Introduction

From a practical perspective, vitamin A is the most important vitamin (McDowell, 2000). Several forms of vitamin A exist, including retinol, the major circulating form; retinyl esters, the main form in which vitamin A is stored by cells; and the activated forms retinoic acid and retinaldehyde (Lespine et al., 1998). Since they are critical to the physiology of the retina, these oxidative states of retinol are all known as retinoids (Engelking, 2015). Retinol is essential for growth, vision, immunity, cellular metabolism, bone development, and reproduction (EFSA, 2013). Recently, retinol has attracted attention for its antioxidant capacities and its role in myogenesis, preadipocyte formation, and the reduction of milk protein allergens (Wang et al., 2018; Hufnagl et al., 2018; Harris et al., 2018; Yu et al., 2022). The main functions of vitamin A and manifestations of deficiency are presented in Table 1.

**Table 1.** Functions and manifestations of vitamin A deficiency

| Chemical form | Functions | Manifestations of deficiency |
|---|---|---|
| Retinol<br>Retinal<br>Retinoic acid | Vision<br>Cell growth and differentiation<br>Immune modulation<br>Reproduction and fetal development<br>Bone health<br>Skin health<br>Gene exprecion<br>Antioxidant capacity<br>Anti-neoplastic properties | Impaired immune function<br>Reproductive and fetal abnormalities<br>Impaired bone development<br>Growth retardation<br>Bitot's spots: Keratomalacia<br>Dry eyes: Xerophthalmia<br>Night blindness: Nyctalopia<br>Skin problems: Hyperkeratosis or follicular hyperkeratosis<br>Anemia |

Preformed types of vitamin A are only found in feed of an animal origin (Carazo et al., 2021). Some compounds in the class of polyisoprenoid plant pigments (carotenoids), may yield retinoids metabolically and, therefore, possess vitamin A activity (Combs and McClung, 2017). However, enzymatic processes can only convert a small percentage of carotenoids into retinoid precursors of retinol or retinoic acid in animals (Solomons, 1994). The extent to which they are utilized depends, among other factors, on the form of binding, animal species, diet composition and vitamin A supply status. Basal feedstuffs, such as cereal grains and roughages, commonly used in ruminant nutrition are deficient in vitamin A (Salinas-Chavira et al., 2014). It is therefore common practice to supplement herbivorous diets with sources of retinol (mostly in the form of retinyl acetate). The importance of providing an adequate amount of this vitamin, no matter how it is administered, has been well recognized (Lespine et al., 1998). Vitamin A requirements, however, depend on a wide range of environmental, genetic, and nutritional factors (Chen et al., 2015).

A comprehensive understanding of vitamin A metabolism in ruminants is essential for accurately assessing dietary requirements, identifying health risks associated with inadequate intake, and informing appropriate supplementation strategies. It is crucial to grasp the complex interplay between vitamin A metabolism, dietary intake, and supplementation for optimal health and wellbeing.

This review provides a thorough evaluation of the metabolism of vitamin A in cattle, including the effects of supplemental retinol on various aspects such as immune function, antioxidant capacity, reproduction, myogenesis, preadipocyte formation, the allergenic potency of milk proteins, and the considerations related to proper vitamin A supply in dairy and beef cattle.

## Metabolism of Vitamin A

Ruminants, such as cattle, have a complex digestive system comprising of several stomach compartments, with the rumen being the largest. This compartment houses a diverse microbial ecosystem that plays a crucial role in feed digestion and fermentation (Castillo-Gonzalez et al., 2014). However, it is also a site of potential vitamin A loss or degradation, which can affect the overall metabolism of retinol in ruminants (Mitchell et al., 1967).

Vitamin A may be present in the paunch as both free forms and esterified forms. Retinyl esters can be cleaved by microorganisms, releasing free retinol (Rode et al., 1990), which can then be converted into other forms such as retinal and retinoic acid through a series of enzymatic reactions carried out by the microflora (Srinivasan and Buys, 2019). As a result, the rumen microbiome can play a critical role in determining the bioavailability and metabolism of vitamin A in cattle.

The metabolism of retinol in herbivores is a complex process that involves multiple organs and tissues, including the intestine and liver. In the small intestine, vitamin A esters are hydrolysed by pancreatic lipases, which release retinol into the intestinal lumen. Vitamin A as a fat-soluble vitamin is predominantly absorbed in the proximal part of the small intestine (Goncalves et al., 2015). After absorption, retinol is re-synthesized into esters in the enterocytes and incorporated into chylomicrons. These chylomicrons are then transported through the lymphatic system and bloodstream to deliver vitamin A to the liver (O'Byrne and Blaner, 2013). Several factors affect the absorption of vitamin A in cattle, such as dietary fat content, intestinal health, and the presence of other micronutrients. Higher dietary fat content can enhance vitamin A absorption, while intestinal diseases can impede it in animals (Reboul, 2013; Amimo et al., 2022).

Once in the hepatic tissue, vitamin A is either stored or transported to other cells in the body. In the liver, vitamin A is esterified with long-chain fatty acids to form retinyl esters, which are stored in hepatic stellate cells (Grumet et al., 2013; Haaker et al., 2020). The liver can store large amounts of retinyl esters, which can be mobilized when needed. Retinol-binding protein (RBP) is the primary carrier protein for vitamin A in the circulation, and it binds to retinol to form a complex that is transported to other tissues (Berry and Noy, 2012). The cellular uptake of retinol is mediated by cells expressing the STRA6 membrane protein, which takes up the retinol-RBP complex (Kelly and von Lintig, 2015). Inside the cell, vitamin A is usually converted into its active form, retinoic acid, through two sequential oxidation reactions catalyzed by retinol dehydrogenases and retinal dehydrogenases (Bchini et al., 2013).

Retinaldehyde can be synthesized from dietary carotenoids, like β-carotene, which are absorbed through simple diffusion and cleaved by dioxygenases in the intestinal mucosal cells, as noted by Hynd (2019). The conversion of retinaldehyde to retinoic acid occurs through a two-step oxidation process, with retinaldehyde dehydrogenase catalyzing the addition of a carboxyl group to the aldehyde group in the first step, followed by aldehyde dehydrogenases converting it to retinoic acid (Belyaeva et al., 2019).

The conversion of carotenoids, such as β-carotene, to vitamin A in cattle is inconsistent probably due to the variable activity of the enzyme responsible for this conversion, β-carotene 15,15'-dioxygenase. While β-carotene has the potential to serve as a source of vitamin A in ruminants, its contribution is hindered by factors such as the highly variable content of β-carotene in feedstuffs, limited and unpredictable conversion rates, low stability, and various animal-related factors. Accurately estimating the amount of vitamin A obtained from β-carotene in cattle diets is therefore challenging. In their publication titled "Meeting the vitamin A requirement: the efficacy and importance of β-carotene in animal species," Green and Fascetti (2016) thoroughly analyzed the effectiveness and significance of β-carotene in various animal species, and the reader is referred to this work for a comprehensive understanding of the topic.

In the animal cells, retinoic acid binds to nuclear receptors, including retinoic acid receptors and retinoid X receptors, to regulate gene expression (Huang et al., 2014). The specific functions of vitamin A in cattle depend on the tissue and cell type, but it is essential for vision, immune function, and reproduction.

Vitamin A plays a crucial role in vision, particularly in the process of dark adaptation, which is the ability of the eye to adjust to low levels of light. The visual pigment in the retina, known as rhodopsin, is composed of a protein called opsin and a molecule called 11-cis-retinal, which is derived from vitamin A (Palczewski, 2014). In bright light, rhodopsin is bleached, and 11-cis-retinal is converted into all-trans-retinal, leading to a decrease in visual sensitivity (Chen et al., 2009). However, in low light, enzymes convert all-trans-retinal back into 11-cis-retinal, which can then recombine with opsin to form rhodopsin, restoring visual sensitivity (Kono et al., 2008). Therefore, a deficiency of vitamin A can lead to impaired dark adaptation, night blindness, and even complete blindness (Debelo et al., 2017). Additionally, vitamin A is also involved in maintaining the health of other parts of the eye, such as the cornea and conjunctiva.

Promoting cellular differentiation is one of the most essential functions of vitamin A in growth. Retinol is necessary for the proper differentiation of cells into specialized tissues and organs during embryonic development and throughout the growing period (Barber et al., 2014). Epithelial tissues, which are found in vital areas such as the skin, respiratory system, urinary tract, and digestive tract, depend on vitamin A for their growth and maintenance (Timoneda et al., 2018).

Apart from its role in cellular differentiation, vitamin A also plays a significant role in the growth and development of bones. Retinol deficiency in cattle can impair bone growth and increase the risk of bone malformations (Hayes et al., 1968). Vitamin A's role in promoting the production and activation of osteoblasts, which are responsible for building new bone tissue, can be attributed, at least in part, to its effect on bone health (Henning et al., 2015). Additionally, retinol is involved in regulating bone resorption, which is the natural process of breaking down old bone tissue and replacing it with new bone tissue (Yee et al., 2021).

Given its essential role in a variety of physiological functions, vitamin A is a fundamental micronutrient, and its insufficiency can result in adverse health outcomes, underscoring the importance of ensuring adequate intake. In the following sections of this scientific review paper, we will provide a comprehensive analysis of the impact of vitamin A supplementation on several physiological functions, such as immune function, antioxidant capacity, reproductive performance, myogenesis and preadipocyte formation, and effect on allergenic potency of milk proteins.

## Effect on Immune Function and Antioxidant Capacity

In 1928, Green and Mellanby discovered the role of vitamin A in immune modulation, naming it the "anti-infective vitamin." The results of subsequent animal studies revealed that vitamin A supplementation improved resistance to gram-negative and gram-positive bacteria, and prevented parasitic infections (Chew, 1987).

The impact of hypovitaminosis A on respiratory infections in young animals is well known. To address this issue, a recent study by McGill et al. (2019) focused on bovine respiratory syncytial virus (BRSV), a pathogen that causes lower respiratory tract disease. The researchers developed a nanovaccine containing BRSV proteins encapsulated in polyanhydride nanoparticles. However, when tested on calves with vitamin A deficiency, the vaccine was ineffective, and the most animals did not receive protection from BRSV challenge (Table 2). Moreover, the study found that acute BRSV infection had an adverse effect on both serum and hepatic retinol levels (Figure 1). These findings highlight the critical role of vitamin A in regulating respiratory mucosal immunity in animals.

**Table 2.** Virus recovered from nasal swabs in BRSV challenged calfs* (McGill et al., 2019)

| Days p.i. | VAS Unvac | VAS BRSV NP | VAD Unvac | VAD BRSV NP |
|---|---|---|---|---|
| Day 0 | 0/6 | 0/13 | 0/6 | 0/12 |
| Day 2 | 2/6 | 1/13 | 3/6 | 4/12 |
| Day 4 | 6/6 | 5/13 | 6/6 | 10/12 |

*VAS=vitamin A sufficient; VAD=vitamin A deficient; Unvac=unvaccinated; BRSV NP=nanovaccine containing BRSV. Nasal swab samples were collected on days 0, 2, 4 and 7 after BRSV challenge. Virus isolation was performed on bovine turbinate cells to determine virus shedding. Results are presented as a ratio of BRSV positive samples/total samples.

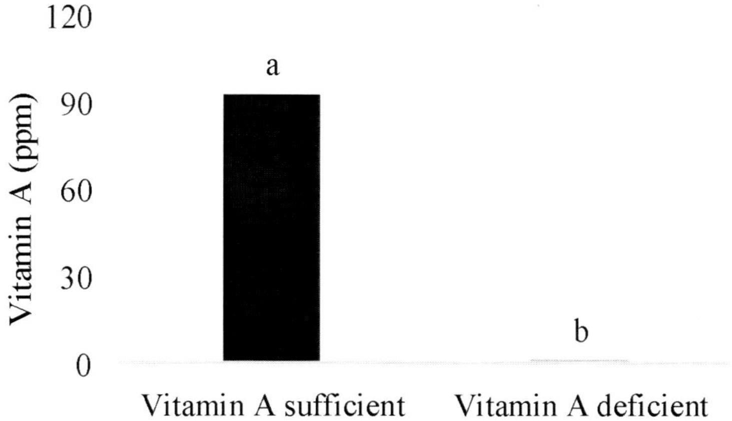


**Figure 1.** Liver retinol level in BRSV challenged calfs (McGill et al., 2019). Liver retinol was measured at necropsy, on day 7 after BRSV infection. [a, b]$P < 0.001$ by student's t-test.

A controlled gene expression appears to enhance both antibody-mediated and cell-mediated immune responses when retinol is provided in an optimal quantity (Oliveira et al., 2018). Jin et al. (2014) found that by supplementing dairy cows with vitamin A at twice the amount recommended by the NASEM (2001), immunoglobulins A, M and G, interleukin-1, glutathione peroxidase, superoxide dismutase and catalase activities were significantly increased along with total antioxidant capacity and the ability to inhibit hydroxyl radicals. In the group supplemented with increased amounts of vitamin A, the authors also observed a significant reduction in the somatic cell count in milk, and a significant reduction in soluble CD8, malonaldehyde and reactive oxygen species.

The prevalence of immune dysregulation is clearly associated with oxidative stress. The immune system uses reactive oxygen species to function, and thus requires adequate levels of antioxidant defences sequentially to prevent the harmful effects of excess free radical production (Victor et al., 2004). It has been shown that fat-soluble vitamins E and A boost immune responses in individuals exposed to specific environmental sources of free radicals (Hajian, 2015). The findings of Jin et al. (2014) suggest that greater vitamin A supplementation (220 IU vs 110 IU/kg BW) in dairy cow diets may improve both antioxidant and immune function. The authors also indicated that adequate vitamin A intake requirement estimates of NASEM (2001) may be insufficient to maintain immunity and antioxidant capacity in dairy cattle.

A study by Ma et al. (2005) found that supplementing beef cattle with 3300 or 4400 IU of vitamin A per kg of DM increased glutathione peroxidase and superoxide dismutase activities in serum significantly more than 1100 IU or 2200 IU of vitamin A per kg of DM. A further increase in vitamin A supplementation levels to 4400 and 5500 IU/kg DM significantly reduced malondialdehyde levels in blood serum. As a result, the authors recommended supplementing beef cattle with 3300 IU vitamin A/kg DM intake. The optimal level of antioxidative protection would, however, require an intake of 5500 IU of vitamin A/kg DM.

As Semba (1999) points out, using supplemental vitamin A should reduce infectious disease morbidity and mortality by enhancing immunity. Among dairy cattle, mastitis is the most common and costly disease in this regard (Halasa et al., 2007). It is often associated with intramammary bacterial infections (Lundberg, 2015). A notable protective role of dietary vitamin A supplementation against mastitis has been demonstrated in ruminant studies (Chew et al., 1982; Chew et al., 1984; Johnson and Chew, 1984; Chew and Johnson, 1985; Van Merris et al., 2004).

By 1982, Chew et al. had already shown that blood plasma levels of vitamin A and β-carotene had a highly significant correlation with California Mastitis Test scores. In the study, dairy cows with low plasma vitamin A and β-carotene levels scored higher than those with higher levels.

Unlike the previously described studies, Oldham et al. (1991) reported no effect of vitamin A on udder health. However, dairy cows supplemented with higher levels of vitamin A (170,000 IU/day) produced significantly more fat-corrected milk than animals receiving only 50,000 IU of vitamin A per day, or 50,000 IU of vitamin A+300 mg β-carotene per day.

More recent studies show that retinoic acid is important for modulating the immune response during acute mastitis, indicating that the metabolism of vitamin A changes profoundly during the course of this disease (Van Merris et al., 2004).

Retinoids, as noted by Semba (1999), influence the immune system in a variety of ways, including humoral immune responses such as the expression of mucin and keratin, cytokines, and adhesion molecules such as ICAM-1. Furthermore, retinoids impact cellular immune responses such as hematopoiesis, neutrophil development, differentiation, and function, natural killer cells, monocytes/macrophages, Langerhans cells, T and B lymphocytes, immune responses resembling T helpers type 1 and 2, immunoglobulin production, and apoptosis.

In a recent study by Agustinho et al. (2021), lactating cows supplemented with vitamin A prepartum had a lower milk somatic cell count than those without vitamin A (Figure 2). The authors attributed this observation to a balanced immune response resulting from retinol supplementation. Consequently, improving both immune and antioxidant functions in ruminants through vitamin A supplementation can be considered essential to health, growth, milk quality and udder antioxidant activity (Jin et al., 2014).

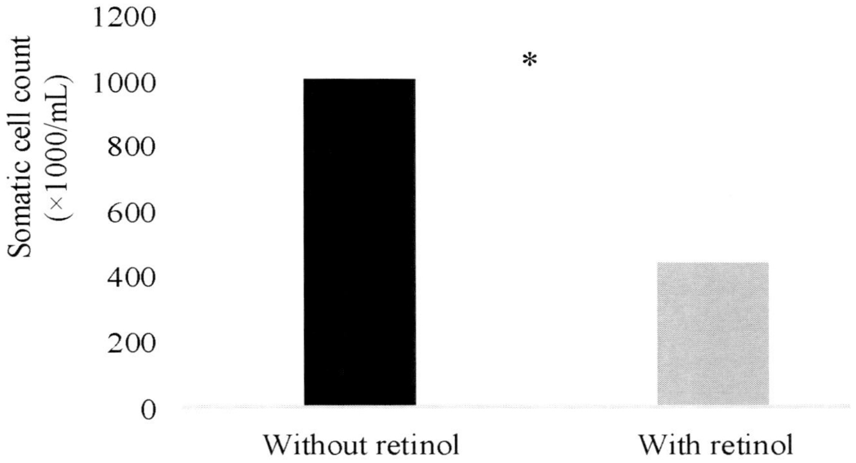


**Figure 2.** Somatic cell count for multiparous Holstein dairy cows (n = 80, total) that were fed diets with or without vitamin A (110 IU/kg body weight or none) during prepartum period (from day -35 to the day of calving; adapted from Agustinho et al., 2021). All cows received a common lactation ration postpartum. *Dietary vitamin A interaction ($P = 0.04$).

## Effect on Reproductive Function

Vitamin A plays a crucial role in maintaining the structure and function of the reproductive tract in both males and females. In males, it is required for the normal development and maturation of spermatozoa, and its deficiency can lead to reduced sperm count and motility (Hogarth and Griswold, 2010). In females, vitamin A is involved in the regulation of follicular development, ovulation, and the maintenance of pregnancy (Clagett-Dame and Knutson, 2011). Its deficiency can cause irregular estrous cycles, embryonic death, and fetal abnormalities (McDowell, 2000). To stimulate the production of testosterone in males, Leydig cells need to differentiate, a process that can be facilitated by retinol. Similarly, retinol also helps to promote the differentiation of theca cells in females, which are responsible for estrogen production (Yang et al., 2018). Therefore, it is important to ensure adequate vitamin A intake in the diet of cattle to maintain optimal reproductive health and productivity.

The study conducted by Shaw et al. (1995) demonstrated that administering vitamin A (retinol palmitate) in beef cows during superovulation had a positive impact on embryo quality. The results showed that the cows treated with vitamin A had a significantly higher ($P < 0.05$) mean number of high-quality and total transferable embryos than the control group that received corn oil alone. Moreover, the mean number of blastocysts was also greater in the cows treated with vitamin A, suggesting an improvement in embryo development. Notably, the administration of vitamin A did not affect ovulation rate or the total number of embryos recovered, indicating that its positive impact was solely on embryo quality. These findings are significant because they suggest that vitamin A supplementation during superovulation in cattle may lead to improved embryo quality, which could ultimately result in higher conception rates and increased productivity. Overall, the study highlights the importance of vitamin A in reproductive performance and the potential benefits of supplementing vitamin A in cattle breeding programs.

It is well established that heat stress during the summer months affects milk yields and fertility in dairy cows (Al-Katanani et al., 1999). It has been observed that when bovine oocytes are exposed to high temperatures, vitamin A has a safeguarding effect (Lawrence et al., 2004).

In their study, Lawrence et al. (2004) investigated the potential protective effects of retinol on bovine oocytes subjected to elevated temperature stress during *in vitro* development. The study found that oocyte development was significantly compromised at 41.0°C, resulting in reduced blastocyst development and fewer total nuclei in derived blastocysts. However, when retinol was administered to the maturation medium, it prevented heat-induced reductions in oocyte development to the blastocyst stage ($P < 0.05$). These results suggest that retinol may have a protective effect on oocytes exposed to heat stress, potentially improving their development and viability. The results of this study are noteworthy in the context of reproductive management in cattle, as heat stress can be a common and challenging issue in many production systems.

*In vitro*, retinol was also found to be beneficial for challenged oocysts, as shown by Livingston et al. (2002). The researchers explored the effects of adding retinol to media containing developing bovine embryos in both low and atmospheric oxygen atmospheres. In the first experiment, bovine oocytes were matured with varying concentrations of retinol, fertilized, and cultured in a modified synthetic oviductal fluid under low oxygen conditions. The results showed that retinol treatment tended to increase blastocyst formation compared to controls. When blastocyst development rates fell below 20% in the control groups, retinol treatment improved embryonic development significantly ($P < 0.02$). In the second and third experiments, the addition of 5 micromolar retinol to the embryo culture medium did not improve cleavage or blastocyst rates under low oxygen conditions, but significantly increased ($P < 0.001$) blastocyst development under atmospheric conditions. These findings suggest that retinol supplementation to the maturation medium can improve embryonic development of bovine oocytes. This positive effect of retinol on stressed oocysts is unclear but Lawrence et al. (2004) questioned whether it might result from its role in the cell's antioxidative system.

Finally, Michal et al. (1994) examined the effect of vitamin A and β-carotene supplementation on the reproductive performance of Holstein cows during the periparturient period. The results showed that cows supplemented with β-carotene had higher concentrations of plasma β-carotene and retinol than unsupplemented cows. β-Carotene and vitamin A supplementation also increased plasma retinol levels and reduced the incidence of retained placenta and metritis ($P < 0.1$), two common reproductive disorders in cows. These findings suggest that dietary supplementation with β-carotene and vitamin A can positively impact the reproductive health of cows during the periparturient period.

Therefore, in dairy cows, it can be concluded that vitamin A promotes the reproductive process and can reduce fertility problems.

## Effect on Myogenesis and Preadipocyte Formation

Vitamin A plays a crucial role in animal growth and development. It appears that retinoic acid signaling has a major impact on myogenic differentiation in mammals (Zhu et al., 2009; Ryan et al., 2012). Wang et al. (2018) reported that vitamin A administration improved cattle growth by increasing PAX7 positive satellite cells and the expression of myogenic marker genes including PAX7, MYF5, MYOD and MYOG. As a result of retinyl palmitate treatment during an early stage, muscle fiber sizes were larger in vitamin A-treated cattle at harvest (Figure 3). The authors suggested that vitamin A levels in beef cattle need to be carefully managed in order to achieve optimal growth performance.

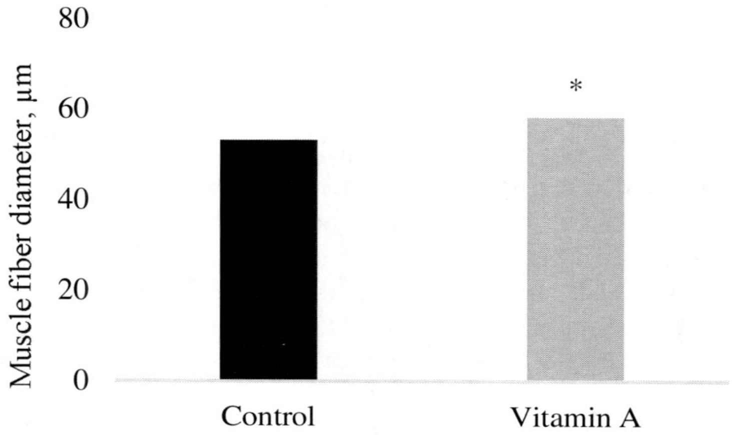


**Figure 3.** Average latissimus dorsi muscle fiber diameter at harvest (adapted from Wang et al., 2018). Angus steer calves were administered with 0 (control) or 150,000 IU vitamin A (retinyl palmitate) per calf at birth and 1 month of age. The resulting steers were harvested at 14 months of age. *$P < 0.05$; Mean, n = 9.

Beef palatability is greatly influenced by intramuscular fat content, or marbling (Moore et al. 2012). The study by Yu et al. (2022) showed that vitamin A/retinoic acid promoted vascular endothelial growth factor A and increased intramuscular adipose progenitors, resulting in the development of adipocytes within muscle tissue. The authors concluded that beef marbling can be promoted by vitamin A when administered at neonatal stage, without effects on overall fat content. Similarly, Harris et al. (2018) found that administering vitamin A at birth resulted in higher weaning weights in beef cattle due to improved muscle growth and enhanced marbling fat production (Table 3). As a result, vitamin A administration during early development could offer beef cattle producers a practical way to enhance marbling and improve beef production efficiency.

**Table 3.** Impacts of neonatal vitamin A administration on cattle growth performance* (adapted from Harris et al., 2018)

| Parameter | 0 IU (n=9) | 150,000 IU (n=7) | 300,000 IU (n=9) | SE |
|---|---|---|---|---|
| Birth to weaning | | | | |
| Birth weight, kg | 35.1 | 35.6 | 35.6 | 0.58 |
| Average daily gain, kg/d | 0.88[b] | 0.98[a] | 1.00[a] | 0.02 |
| Backgrounding | | | | |
| Weaning weight at d 210, kg | 219.2 | 248.7 | 246.0 | 5.98 |
| Gain:Feed ration, kg | 0.149 | 0.153 | 0.156 | 0.031 |
| Finishing | | | | |
| Weight at 308 d, kg | 312.1[b] | 333.0[ab] | 339.7[a] | 8.65 |
| Feed:Gain ration, kg | 4.35 | 4.76 | 4.79 | 0.13 |

*Angus steer calves (n = 30) were randomly allocated to three treatment groups at birth, receiving 0, 150,000, or 300,000 IU of vitamin A at both birth and one month of age. [a,b] Mean values within a row with no common superscript differ significantly ($P < 0.05$).

## Effect on Allergenic Potency of Milk Proteins

Cow's milk is one of the most common causes of food allergies in children and can also develop in adults. It is usually caused by bovine β-lactoglobulin (Selo et al., 1999; Hochwallner et al. 2014). This protein has the ability to bind a variety of hydrophobic molecules, for example retinoic acid (Konuma et al., 2007). In a recent study conducted by Hufnagl et al. (2018), retinoic acid was shown to inhibit the immunogenicity of β -lactoglobulin by binding to, and masking, its main cell epitope. According to the authors, feeding vitamin A to cows can enhance the loading of β-lactoglobulin with retinoic acid, thereby reducing the development of allergies to this component of cow's milk. Further research is warranted to assess whether such a strategy would be effective and feasible.

## Guidance for Proper Vitamin A Supplementation

There are insufficient data to establish definitive vitamin A requirements for cattle, therefore "Adequate Intake" levels have been set by the NASEM (2000, 2001) and NASEM (2016, 2021). In ruminant nutrition, vitamin A supplementation must be considered for the prevention of deficiency symptoms, and to support optimum health, reproduction and performance. As Weiss stated (2017), the absence of a deficiency does not imply an adequate dietary vitamin A level. Typically, healthy immune function requires higher vitamin A levels than those related to performance or reproduction (Weiss, 1998). Considering the benefits, the severe degradation of vitamin A in the rumen, and very high tolerability (up to 1.3 million IU vitamin A/d), feeding twice the adequate vitamin A intake requirement estimates of NASEM (2001) is justifiable (Jin et al., 2014).

According to the new NASEM (2021) recommendations, dry cows, growing heifers and lactating animals giving less than 35 kg of milk per day do not need more vitamin A than the value estimated by the NASEM (2001). The adequate retinol intake for animals producing more than 35 kg milk per day is now calculated as follows: $110 \times BW + 1000 \times (milk - 35)$. Thus, to replace the retinol secreted in milk by high performing dairy cows, an additional 1000 IU of vitamin A must be supplemented for each kilogram of milk. Despite this adjustment, nutritionists usually only consider the NASEM estimates for vitamins to be the minimum levels required to prevent clinical deficiencies (McDowell, 1989; Kırkpınar and Açıkgöz, 2018). According to McDowell (2006), however, the optimum vitamin supplementation level is the quantity that achieves the best growth rate, feed utilization, and health (including immune competency), while also providing adequate body reserves.

Finally, antioxidant capacity must be considered. To produce one liter of milk, 540 liters of blood are perfused into the udder (Potapow, 2010). Consequently, the higher the milk yield, the more oxidative stress occurs in the mammary glands (Bae et al., 2017). It therefore seems logical to give high-performing dairy cows more fat-soluble antioxidant vitamins such as retinol and α-tocopherol due to the increased redox imbalance. Vitamin A supplementation of 1000 IU per additional liter of milk (animals with >35 L/head/d) is unlikely to compensate for the increased oxidative stress in high-performing dairy cows. Similarly, modern beef cattle breeds, which are genetically enhanced to produce fast growth and meat deposition, exhibit increased metabolic turnover and oxidative stress. Retinol supplementation above the levels defined as adequate by NASEM (2016; 2021) (Tables 4 and 5) is required by high-performing animals under stress to maximize antioxidant and immune function (Ma et al., 2005; Jin et al., 2014). Furthermore, supplemental retinyl acetate may sustain significant losses during storage and processing, which cannot be accounted for in scientific vitamin norms (Hirai et al., 2023).

To close this gap and finally establish vitamin A requirements for cattle, high-quality dose-response trials are needed. These are costly and require a significant number of animals, making studies of large ruminants difficult. It is nevertheless imperative in the years to come that considerable effort is made to understand retinol requirements in dairy and fattening cattle at all stages of production.

**Table 4.** Estimated adequate vitamin A intake in dairy cattle, NASEM (2001) and NASEM (2021)

| Committee | Adequate intake of vitamin A, IU/kg BW | | | |
|---|---|---|---|---|
| | Growing heifers | Dry cows | Lactating cows, (<35 kg milk/d) | Lactating cows, (>35 kg milk/d) |
| NASEM 2001 | 110 | 110 | 110 | n/a |
| NASEM (2021) | 110 | 110 | 110 | 110+(1000×(milk-35)) |

**Table 5.** Estimated adequate vitamin A intake in beef cattle, NASEM (2000) and NASEM (2016)

| Committee | Adequate intake of vitamin A, IU/kg BW | | |
|---|---|---|---|
| | Growing and finishing | Gestating cows | Lactating cows |
| NASEM 2000 | 2200 | 2800 | 3900 |
| NASEM 2016 | 2200 | 2800 | 3900 |

Ultimately, balancing vitamin A supply for ruminants involves formulating a diet that provides sufficient vitamin A to meet the animal's requirements, without exceeding safe levels. To ensure the safe and appropriate use of retinyl esters in cattle farming, it is essential to follow specific guidelines that limit the supplemental levels of vitamin A to the recommended intake levels as established by relevant authorities such as EC in the EU or CFIA in Canada. Any use of vitamin A in ruminants should be carried out within the framework of legal provisions and guidelines to prevent over-supplementation and potential negative impacts on animal health and welfare. Adherence to these guidelines is crucial to ensure that vitamin A supplementation in cattle farming is sound and effective in promoting optimal health and productivity. Hypervitaminosis A is highly unlikely to occur in cattle due to the strict regulation of vitamin A levels in animal feed, and it has not been reported in practical farming or husbandry scenarios, although it can be induced in certain experimental conditions.

## Conclusion

Cattle require retinol as a key vitamin. It is common for ruminant feeds to be deficient in vitamin A. Therefore, synthetic retinyl acetate supplementation is given to the animals to meet their dietary requirements. The positive effects of retinol supplementation and injection have been demonstrated on cattle reproduction, performance, immunity, and health. Balanced cattle diets can maximize performance under specific conditions, if we know when it is appropriate, or best, to administer greater retinol supplementation or injection during production, and consider its effects on other vitamins, especially vitamins E and D. Therefore, further research is needed on the interaction of retinol with other fat-soluble bioactive agents. Additionally, we advocate establishing retinol requirements which take into account the function of vitamin A, genetics, and cattle farming methods based on the current state of science.

## Acknowledgments

Thanks to Ute Obermueller-Jevic, PhD for her constructive comments on the draft manuscript.

## Disclaimer

None.

## Conflicts of Interest

The authors have no conflicts of interest to declare.

## Declaration of Funding

The authors reported there is no funding associated with the work featured in this article.

## References

Agustinho BC, Ramsey KC, Moore C, Tsai CY, Scholte CM, McGuire MA, Rezamand P. Effect of different combinations of dietary vitamin a, protein levels, and monensin on inflammatory markers and metabolites, retinol-binding protein, and retinoid status in periparturient dairy cows. *Animals* (2021) 11:2605.

Al-Katanani Yaser M, Webb DW, Hansen PJ. Factors affecting seasonal variation in 90-day nonreturn rate to first service in lactating holstein cows in a hot climate. *Journal of Dairy Science* (1999) 82:2611-2616.

Amimo JO, Michael H, Chepngeno J, Raev SA, Saif LJ, Vlasova AN. Immune impairment associated with vitamin A deficiency: insights from clinical studies and animal model research. *Nutrients* (2022) 14:5038.

Bae H, Jeong C, Cheng W, Hong K, Seo H, Han S. Oxidative stress-induced inflammatory responses and effects of N-acetylcysteine in bovine mammary alveolar cells. *Journal of Dairy Research* (2017) 84:418-425.

Barber T, Esteban-Pretel G, Marin MP, Timoneda J. Vitamin A deficiency and alterations in the extracellular matrix. *Nutrients* (2014) 6(11):4984-5017.

Bchini R, Vasiliou V, Branlant G, Talfournier F, Rahuel-Clermont S. Retinoic acid biosynthesis catalyzed by retinal dehydrogenases relies on a rate-limiting conformational transition associated with substrate recognition. *Chemico-Biological Interactions* (2013) 202:78-84.

Belyaeva OV, Adams MK, Popov KM, Kedishvili NY. Generation of retinaldehyde for retinoic acid biosynthesis. *Biomolecules* (2019) 10:5.

Berry DC, Noy N. Signaling by vitamin A and retinol-binding protein in regulation of insulin responses and lipid homeostasis. *Biochimica et Biophysica Acta* (2012) 1821(1):168-76.

Carazo A, Macákova K, Matoušová K, Kujovská Krčmová L, Protti M, Mladěnka P. Vitamin A update: Forms, sources, kinetics, detection, function, deficiency, therapeutic use and toxicity. *Nutrients* (2021) 13:1-36.

Castillo-Gonzalez AR, Burrola-Barraza ME, Dominguez-Viveros J, Chávez-Martínezb A. Rumen microorganisms and fermentation. *Archivos de Medicina Veterinaria* (2014) 46(3):349-361.

Chen C, Blakeley LR, Koutalos Y. Formation of all-trans retinol after visual pigment bleaching in mouse photoreceptors. *Investigative Ophthalmology & Visual Science* (2009) 50(8):3589-95.

Chen F, Jiang Z, Jiang S, Li L, Lin X, Gou Z, Fan Q. Dietary vitamin A supplementation improved reproductive performance by regulating ovarian expression of hormone receptors, caspase-3 and Fas in broiler breeders. *Poultry Science* (2015) 95:30-40.

Chew BP, Hollen LL, Hillers JK, Herlugson ML. Relationship between vitamin A and β-carotene in blood plasma and milk and mastitis in Holsteins. *Journal of Dairy Science* (1982) 65:2111.

Chew BP, Holpuch DM, O'Fallon JV. Relative concentrations of vitamin A and β-carotene in bovine and porcine plasma, liver, corpora lutea and follicular fluid. *Journal of Dairy Science* (1984) 67:1316.

Chew BP, Johnston LA. Effects of supplements vitamin A and β-carotene on mastitis in dairy cows. *Journal of Dairy Science* (1985) 68 (Suppl. 1):191.

Chew BP. Vitamin A and ß-carotene on host defence. Symposium:Immune function:Relationship of changes nutrition and disease control. *Journal of Dairy Science* (1987) 70:2732.

Clagett-Dame M, Knutson D. Vitamin A in reproduction and development. *Nutrients* (2011) 3(4):385-428.

Combs GF, McClung JP. *The Vitamins*. London, UK: Academic Press; 2017.

Debelo H, Novotny JA, Ferruzzi MG. Vitamin A. *Advances in Nutrition* (2017) 8:992-994.

EFSA (European Food Safety Authority). Scientific Opinion on the safety and efficacy of vitamin A (retinyl acetate, retinyl palmitate and retinyl propionate) as a feed additive for all animal species and categories. *EFSA Journal* (2013) 11:3037.

Engelking LR. *Textbook of Veterinary Physiological Chemistry*. London, UK: Academic press; 2015.

Goncalves A, Roi S, Nowicki M, Dhaussy A, Huertas A, Amiot M-J, Reboul E. Fat-soluble vitamin intestinal absorption: Absorption sites in the intestine and interactions for absorption. *Food Chemistry* (2015) 172C:155-60.

Green AS, Fascetti AJ. Meeting the vitamin A requirement: The efficacy and importance of β-carotene in animal species. *The Scientific World Journal* (2016):7393620. doi: 10.1155/2016/7393620.

Green HN, Mellanby E. Vitamin A as an anti-infective agent. *British Medical Journal* (1928) **2**:691.

Grumet L, Taschler U, Lass A. Hepatic retinyl ester hydrolases and the mobilization of retinyl ester stores. *Nutrients* (2016) 9(1):13.

Haaker MW, Vaandrager AB, Helms JB. Retinoids in health and disease: A role for hepatic stellate cells in affecting retinoid levels. *Biochimica et Biophysica Acta - Molecular & Cell Biology of Lipids* (2020), 1865 (6):158674.

Hajian S. Positive Effect of Antioxidants on Immune System. *Immunopathologia Persa* (2015) 1:e02.

Halasa T, Huijps K, Østerås O, Hogeveen H. Economic effects of bovine mastitis and mastitis management: A review. *Veterinary Quarterly* (2007) 29:18-31.

Harris CL, Wang B, Deavila JM, Busboom JR, Maquivar M, Parish SM, McCann B, Nelson ML, Du M. Vitamin A administration at birth promotes calf growth and intramuscular fat development in Angus beef cattle. *Journal of Animal Science and Biotechnology* (2018) **9**:55.

Hayes KC, Nielsen SW, Eaton HD. Pathogenesis of the optic nerve lesion in vitamin a-deficient calves. *Archives of Ophthalmology* (1968) 80(6):777-787.

Henning P, Conaway HH, Lerner UH. Retinoid receptors in bone and their role in bone remodeling. *Frontiers in Endocrinology* (Lausanne) (2015) 6:31.

Hirai RA, De Leon D, Randig-Biar M, Silva A, Sanchez E, McElroy AP, Bailey CA, Martinez N, Sokale A, Music L. Evaluation of the stability of vitamin A acetate concentrates mixed in a vitamin-trace mineral premix over a 56-day high temperature and humidity storage stress. *The International Poultry Scientific Forum* (2023), Atlanta, USA.

Hochwallner H, Schulmeister U, Swoboda I, Spitzauer S, Valenta R. Cow's milk allergy: from allergens to new forms of diagnosis, therapy and prevention. *Methods* (2014) 66:22-33.

Hogarth CA, Griswold MD. The key role of vitamin A in spermatogenesis. *The Journal of Clinical Investigation* (2010) 120(4):956-62.

Huang P, Chandra V, Rastinejad F. Retinoic acid actions through mammalian nuclear receptors. *Chemical Reviews* (2014) 114(1):233-54.

Hufnagl K, Ghosh D, Wagner S, Fiocchi A, Dahdah L, Bianchini R, Braun N, Steinborn R, Hofer M, Blaschitz M, Roth GA, Hofstetter G, Roth-Walter F, Pacios LF, Jensen-Jarolim E. Retinoic acid prevents immunogenicity of milk lipocalin Bos d 5 through binding to its immunodominant T-cell epitope. *Scientific Reports* (2018) 8:1598.

Hynd PI. *Animal nutrition: From theory to practice*. Clayton South, Australia: CSIRO Publishing; 2019.

Jin L, Yan S, Shi B, Bao H, Gong J, Guo X, Li J. Effects of vitamin A on the milk performance, antioxidant functions and immune functions of dairy cows. *Animal Feed Science & Technology* (2014) 192:15-23.

Johnston LA, Chew BP. Relationship between peripanum plasma vitamin A, β-carotene and mastitis in dairy cows. *Journal of Dairy Science* (1984) 67:1832.

Kelly M, von Lintig J. STRA6: Role in cellular retinol uptake and efflux. *The Hepatobiliary Surgery & Nutrition* (2015) 4 (4):229-42.

Kırkpınar F, Açıkgöz Z. *Feeding: Animal Husbandry and Nutrition*. Available from: http://dx.doi.org/10.5772/intechopen.69938; 2018.

Kono M, Goletz PW, Crouch RK. 11-cis- and all-trans-retinols can activate rod opsin: rational design of the visual cycle. *Biochemistry* (2018) 47(28):7567-71.

Konuma T, Sakurai K, Goto Y. Promiscuous binding of ligands byβ-lactoglobulin involves hydrophobic interactions and plasticity. *Journal of Molecular Biology* (2007) 368:209-218.

Lawrence JL, Payton RR, Godkin JD, Saxton AM, Schrick FN, Edwards JL. Retinol improves development of bovine oocytes compromised by heat stress during maturation. *Journal of Dairy Science* (2004) 87:2449-2454.

Lespine A, Periquet B, Garcia J, Ghisolfi J, Thouvenot JP. Retinol and retinyl ester concentrations in rat tissues during total parenteral nutrition. *The Journal of Nutritional Biochemistry* (1998) 9:316-323.

Livingston TE, Eberhardt DM, Godkin JD. The effects of retinol on bovine embryo development in vitro. *Biology of Reproduction* (2002) 66:104-105.

Lundberg A. *Mastitis in dairy cows. Genotypes, spread, and infection outcome of three important udder pathogens*. Doctoral thesis No. 2015:28, Uppsala, Sweden: Swedish University of Agricultural Sciences; 2015.

Ma XM, Yang ZB, Yang WR, Song ML. Effect of different vitamin A levels in diets on antioxidant ability of beef cattle. *Chinese Journal of Animal Nutrition* (2005) 17:31-35.

McDowell LR. Vitamin nutrition of livestock animals: Overview from vitamin discovery to today. *Canadian Journal of Animal Science* (2006) 86:171-179.

McDowell LR. *Vitamins in Animal and Human Nutrition*. USA: Iowa State University Press; 2000.

McDowell LR. *Vitamins in animal nutrition: Comparative aspects to human nutrition*. USA: Academic Press Inc; 1989.

McGill JL, Kelly SM, Guerra-Maupome M, Winkley E, Henningson J, Narasimhan B, Sacco RE. Vitamin A deficiency impairs the immune response to intranasal vaccination and RSV infection in neonatal calves. *Scientific Reports* (2019) 9(1):15157.

Michal JJ, Heirman LR, Wong TS, Chew BP, Frigg M, Volker L. Modulatory effects of dietary β-carotene on blood and mammary leukocyte function in periparturient dairy cows. *Journal of Dairy Science* (1994) 77:1408-1421.

Mitchell GE, Jr, Little CO, Hayes BW. Pre-intestinal destruction of vitamin A by ruminants fed nitrate. *Journal of Animal Science* (1967) 26(4):827-829.

Moore MC, Gray GD, Hale DS, Kerth CR, Griffin DB, Savell JW, O'Connor ME. National Beef Quality Audit-2011: In-plant survey of targeted carcass characteristics related to quality, quantity, value, and marketing of fed steers and heifers. *Journal of Animal Science* (2012) 90:5143-5151.

NASEM (National Academies of Sciences, Engineering, and Medicine). *Nutrient requirements of dairy cattle*. Washington, DC: The National Academies Press; 2021.

NASEM (National Academies of Sciences, Engineering, and Medicine). *Nutrient requirements of beef cattle*. Washington, DC: The National Academies Press; 2016.

NASEM (National Academies of Sciences, Engineering, and Medicine). *Nutrient requirements of beef cattle*. Washington, DC: National Academy Press; 2000.

NASEM (National Academies of Sciences, Engineering, and Medicine). *Nutrient requirements of dairy cattle*. Washington, DC, USA: National Academic Science; 2001.

O'Byrne SM, Blaner WS. Retinol and retinyl esters: biochemistry and physiology. *Journal of Lipid Research* (2013) 54 (7):1731-43.

Oldham ER, Eberhart RJ, Muller LD. Effects of supplemental vatamin A and β-caroteneduring the dry period and early lactation on udder health. *Journal of Dairy Science* (1991) 74:1345-1352.

Oliveira LM, Teixeira FME, Sato MN. Impact of retinoic acid on immune cells and inflammatory diseases. *Mediators of Inflammation* (2018):doi: 10.1155/2018/3067126.

Palczewski K. Chemistry and biology of the initial steps in vision: the Friedenwald lecture. *Investigative Ophthalmology & Visual Science* (2014) 55 (10):6651-72.

Potapov A. *Investigation of mammary blood flow changes by transrectal colour Doppler sonography in an Escherichia coli mastitis model*. PhD Thesis, Munich, Germany; 2010.

Reboul E. Absorption of vitamin A and carotenoids by the enterocyte: focus on transport proteins. *Nutrients* (2013) 5 (9):3563-3581.

Rode LM, McAllister TA, Cheng K-J. Microbial degradation of vitamin a in rumen fluid from steers fed concentrate, hay or straw diets. *Canadian Journal of Animal Science* (1990) 70:227-233.

Ryan T, Liu J, Chu A, Wang L, Blais A, Skerjanc IS. Retinoic acid enhances skeletal myogenesis in human embryonic stem cells by expanding the premyogenic progenitor population. *Stem Cell Reviews & Reports* (2012) 8:482-493.

Salinas-Chavira J, Arrizon AA, Barreras A, Chen CZ, Plascencia A, Zinn RA. Evaluation of supplemental vitamin A and E on 56-day growth performance, dietary net energy, and plasma retinol and tocopherol concentrations in Holstein steer calves. *The Professional Animal Scientist* (2014) 30:510-514.

Selo I, Clement G, Bernard H, Chatel JM, Creminon C, Peltre G, Wal JM. Allergy to bovine ß--lactoglobulin: specificity of human IgE to tryptic peptides. *Clinical & Experimental Allergy* (1999) 29:1055-1063.

Semba RD. Vitamin A and immunity to viral, bacterial and protozoan infections *Proceedings of the Nutrition Society* (1999) 58:719-727.

Shaw DW, Farin PW, Washburn SP, Britt JH. Effect of retinol palmitate on ovulation rate and embryo quality in superovulated cattle. *Theriogenology* (1995) 44:51-58.

Solomons NW. *Dairy products in human health and nutrition*. Netherlands: Taylor & Francis; 1994.

Srinivasan K, Buys EM. Insights into the role of bacteria in vitamin A biosynthesis: Future research opportunities. *Critical Reviews in Food Science & Nutrition* (2019) 59(19):3211-3226.

Timoneda J, Rodríguez-Fernández L, Zaragozá R, Marín MP, Cabezuelo MT, Torres L, Viña JR, Barber T. Vitamin A deficiency and the lung. *Nutrients* (2018) 10 (9):1132.

Van Merris V, Meyer E, Duchateau L, Blum J, Burvenich C. All-trans retinoic acid is increased in the acute phase-related hyporetinemia during escherichia coli mastitis. *Journal of Dairy Science* (2004) 87:980-987.

Victor VM, Rocha M, De la Fuente M. Immune cells: Free radicals and antioxidants in sepsis. *International Immunopharmacology* (2004) 4:327-347.

Wang B, Nie W, Fu X, M de Avila J, Ma Y, Zhu MJ, Maquivar M, Parish SM, Busboom JR, Nelson ML, Du M. Neonatal vitamin A injection promotes cattle muscle growth and increases oxidative muscle fibers. *Journal of Animal Science & Biotechnology* (2018) 9:82.

Weiss WP. A 100-Year Review: From ascorbic acid to zinc-mineral and vitamin nutrition of dairy cows. *Journal of Dairy Science* (2017) 100:10045-10060.

Weiss WP. Requirements of fat-soluble vitamins for dairy cows: A review. *Journal of Dairy Science* (1998) 81:2493-2501.

Yang Y, Luo J, Yu D, Zhang T, Lin Q, Li Q, Wu X, Su Z, Zhang Q, Xiang Q and Huang Y. Vitamin A promotes leydig cell differentiation via alcohol dehydrogenase. *Frontiers in Endocrinology* (2018) 9:644.

Yee MMF, Chin K-Y, Ima-Nirwana S, Wong SK. Vitamin A and bone health: a review on current evidence. *Molecules* (2021) 26 (6):1757.

Yu X, Ma Y, Luo Y, Tang G, Jiang Z, Zhang J, Ye B, Huang Z, Luo Y, Du M, Wang B. Neonatal vitamin A administration increases intramuscular fat by promoting angiogenesis and preadipocyte formation. *Meat Science* (2022) 191: https://doi.org/10.1016/j.meatsci.2022.108847.

Zhu GH, Huang J, Bi Y, Su Y, Tang Y, He BC, He Y, Luo J, Wang Y, Chen L, Zuo GW, Jiang W, Luo Q, Shen J, Liu B, Zhang WL, Shi Q, Zhang BQ, Kang Q, Zhu J, Tian J, Luu HH, Haydon RC, Chen Y, He TC. Activation of RXR and RAR signaling promotes myogenic differentiation of myoblastic C2C12 cells. *Differentiation* (2009) 78:195-204.

## Chapter 3

# The Influence of Geochemical Characteristics and Geological Formations on Mammal Density and Distribution: A Case Study from Gorny Altai, Russia

**Alexander V. Shitov[1,*], PhD**
**Andrey V. Karanin[1], PhD**
**and Rachel A. Grant[2], PhD**
[1]Gorno-Altaisk State University, Republik of Altai, Russia
[2]Department of Applied Sciences, London South Bank University, London, UK

## Abstract

There has been little research on how the composition of underlying rock formation affects animal species' distribution. The subject is worthy of consideration as it has been shown that ultrabasic and serpentine rocks in particular can give rise to plant biodiversity hotspots with a high level of endemism, and should be prioritized for conservation efforts. Corresponding studies of fauna are lacking. Here we present a comparative analysis of the abundance of mammals and its relationship with geological composition in the area of Gorny Altai, a mountainous region in Russia.

The study reveals significant correlations between underlying geology and the abundance of some species of mammal, even when other factors such as soil type, climate and vegetation are held constant. Intrusive rocks were found to have the greatest impact on variation in mammal distribution whereas sedimentary and metamorphic rocks have

[*] Corresponding Author's Email: sav103@yandex.ru.

In: Advances in Animal Science and Zoology. Volume 21
Editor: Owen P. Jenkins
ISBN: 979-8-89113-291-7

almost no effect. A characteristic feature of magmatic formations is their clear geochemical specialization, i.e., certain geochemical anomalies (Fe, Cu, Au, Hg, Ag, etc.) are confined to intrusions. We suggest that geophysical fields (magnetic and electric fields) and geochemical anomalies associated with intrusive rocks may have an impact on the distribution and species composition of mammals, as well as geodynamic processes such as fault activity.

**Keywords**: mammals, Altai Mountains, geological formations, geochemical characteristics

## Introduction

Geoecology, the study of interactions between the geosphere and ecosystems is a field of study which is under-represented in the current literature. Although there have been several studies looking at the effects of the geosphere on plant communities, similar studies on animal distribution are sparse. The ecology of serpentine soils derived from rocks high in ferromagnesian, or mafic minerals (Kruckeberg, 2002) has been relatively well characterized. Serpentine soils often give rise to highly specialized endemic species of vascular plants and very high biodiversity (Stevanović et al., 2003; Brady et al., 2005; Casazza at al, 2005), making them noteworthy for conservation purposes. It has been reported that in the Mediterranean, benthic communities are species-poor where the rocks have a high quartz content and differences in fish assemblages have been found between limestone and granite (Cattaneo-Vietti et al., 2002; Guidetti & Cattaneo-Vietti 2002; Guidetti et al., 2004).

Geophysical factors which could influence ecosystems include geochemical and geophysical anomalies, inhomogeneity of the Earth's crust: fault zones, accumulations of metal ores, underground water, changes in the exertion of rock, etc. These anomalies are of interest because in the case of geodynamic processes, such as degassing from faults animals are not necessarily adapted to them because of the rarity of their occurrence (Osipov, 1993; Kosinova et al., 2009; Nikolaev, 2014). However, in the case of longstanding or more frequent anomalies, animals will have co-evolved with the geological landscape (Kirschvink et al, 1985). The effects of the geological environment on public health (medical geology) are well documented (Selinus et al., 2005; Finkelman et al., 2007; Shitov, 2010, 2012; Rapant & Cvečkova

2014), so it is logical that animals could be similarly affected by both transitory and longterm geological effects.

In this research we carry out a correlational analysis using mammal abundance data obtained in earlier research and the results of quantitative analysis of the geochemical and geological composition of landscape records in the south-eastern Altai. Long term monitoring of the abundance, species composition and distribution of mammals was carried out as part of a zoological project run by The Institute of Systematics and Ecology of the Altai Republic, which provided the data used in this study.

This former analysis assessed the level of correlation between geological and geochemical landscape variables with quantitative characteristics of animal distribution. Mammal abundance would be expected to vary according to geochemical factors (such as soil types and soil minerals) but the prior analysis showed that correlation coefficients (Kendall's Tau) between mammal abundance and geological data were double that of the correlation between mammal abundance and geochemical data. Consequently, we concluded that the influence of geological aspects of the landscapes on the distribution of mammals in the south-eastern Altai is higher than the comparable influence of the chemical composition of soils of these landscapes (Karanin & Shitov, 2006). Based on the results of this smaller study we carried out a similar correlational analysis on a much larger scale, throughout the Altai Mountains.

## Methodology

### The Previous Analysis

Data on the distribution of mammals in the landscapes of Gorny Altai has been previously collected by the Department of Zoology, Ecology and Genetics of Gorno-Altaisk State University and by Yu. P. Malkov (Table S1; Malkov & Belikov 2005). The records are the result of extensive monitoring of mammal distribution from numerous sources over many years. Data collection methods are shown in the supplementary material Table T2, but are outlined briefly here.

Malkov & Belikov (1995) extrapolated collected data to estimate the number and biomass of mammals in each landscape area. In this earlier study, the biomass of each species was calculated as the product of the number of each individual of that species multiplied by the average weight of the animal. Total mammal biomass was calculated by the sum of species' biomass. Average mammalian masses were calculated from data on the abundance and average masses of each species based on the concept of K - and K-selection (see Begon, Townsend, Harper, 2006). To calculate biodiversity in the South-Eastern Altai the Simpson biodiversity index was used as the main index, and the Shannon biodiversity index as an auxiliary (Karanin, 2004). Additional indicators of community uniformity (uniformity of distribution of species) on Simpson and on Shannon (on density and biomass) were used. For the current analysis, extrapolated animal numbers / $km^2$ were used.

## The Study Site and Its Characteristics

Gorny Altai is a region of Russia, located in the south of Western Siberia, bordering China and Kazakhstan (Coordinates: 82-90°E, 49-52.5°N) (Figure 1). The area is characterized by a complex system of ranges, deep river valleys and wide mountainous troughs. Gorny Altai, which in turn is structurally included in the Western part of the Altai-Sayan folded region, is composed of a mosaic of blocks representing different age geosynclinal-folded systems (Dobretsov et al., 1993; Buslov et al, 1998).

The territory of the Altai Mountains is a complex folded system formed by Precambrian and Paleozoic strata intensively deployed in the Caledonian era of tectogenesis and the Hercynian era of tectogenesis.

In the post-Paleozoic, mountain-folded structures were destroyed and turned into a denudation plain (peneplain). Anticlinoria Gorny Altai (Holynska-Chuyskiy, talitskiy, etc.) is mainly composed of terrigenous flysch series of the upper Cambrian – lower Ordovician overlying Vendian-Cambrian ophiolites, siliceous-shale formation and presumably Precambrian metamorphic rocks, sometimes exposed to the surface.

Superimposed troughs and grabens (the largest – Korgonsky) is made Moloss middle Ordovician – lower Silurian and early Devonian. The deposits are broken by late Devonian granites. Within the Altai Mountains, which has a Caledonian Foundation, widespread volcanic rocks of the plutonic Association of the middle Devonian-early Carboniferous and late Paleozoic granitoids. In Oligocene-Quaternary time, the territory experienced uplift associated with regional compression of the earth's crust caused by the convergence of its limiting lithospheric microplates (Jungar, Tuvan-Mongolian).

The formation of the mountain structure took place on the type of a large arch, which in the last stages of development was deformed by a system of breaks, resulting in a series of block morphostructures in the form of high ridges and troughs separating them in the Central and southern parts. Instrumental observations record vertical movements of the earth's crust, the speed of which reaches several centimeters per year. Uplifts occur unevenly, accompanied by thrusts, which causes the asymmetry of the ridges.

Landscape structure in the provinces of the Altai Mountains varies, the Atlas of the Altai Krai (1978) (Samoilova, 1973) was used as the cartographic standard. Figure 1 shows the study area divided into landscape provinces, which was further subdivided into 26 "landscape-complexes" for the purposes of this study. There were five landscape provinces; North, Northwest, Northeast, Central, East and Southeast. A geological map was superimposed on to this map showing various intrusive and non-intrusive formations.

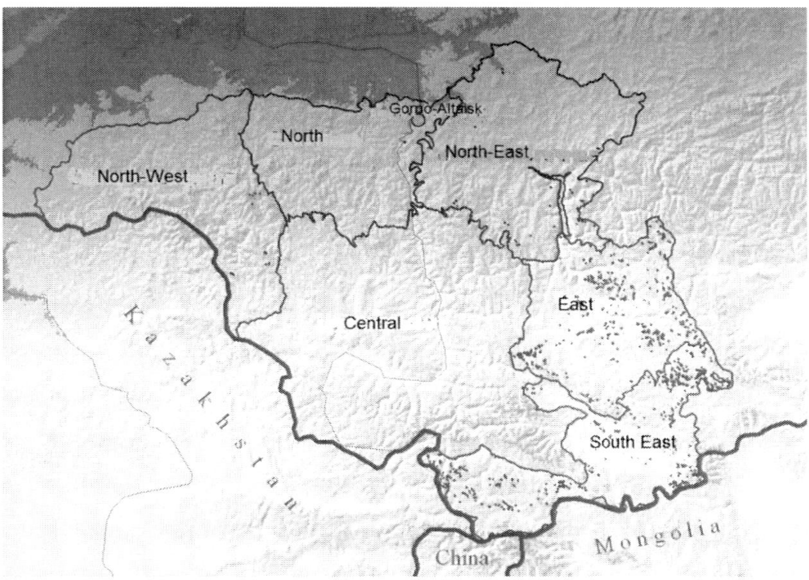


**Figure 1**. Location of landscape provinces of Gorny Altai.

Animal abundance (individuals / km$^2$) was calculated for each landscape-complex, making the landscape-complex the unit of measurement. We used GIS to overlay geological and landscape maps. Since the concept of landscape-complex includes a number of factors (such as altitude, vegetation, soil type and climate), the use of landscape-complexes as a basis allowed us to account for their combined influence. Thus, the combination of landscape and geological maps, allowed us to isolate the influence of geological factors from the sum of other factors that are already taken into account in the landscape – complex unit. Any variation in animal distribution could therefore be assumed to result directly from underlying geology as other factors has been held constant.

A digitized landscape map was used to visualize the distribution of species across the territory (Atlas of the Altai Krai, 1976). Data on the distribution of each mammal species was merged with the landscape database. Figure 2 shows an example of the map produced for *Canis lupus* (wolf).

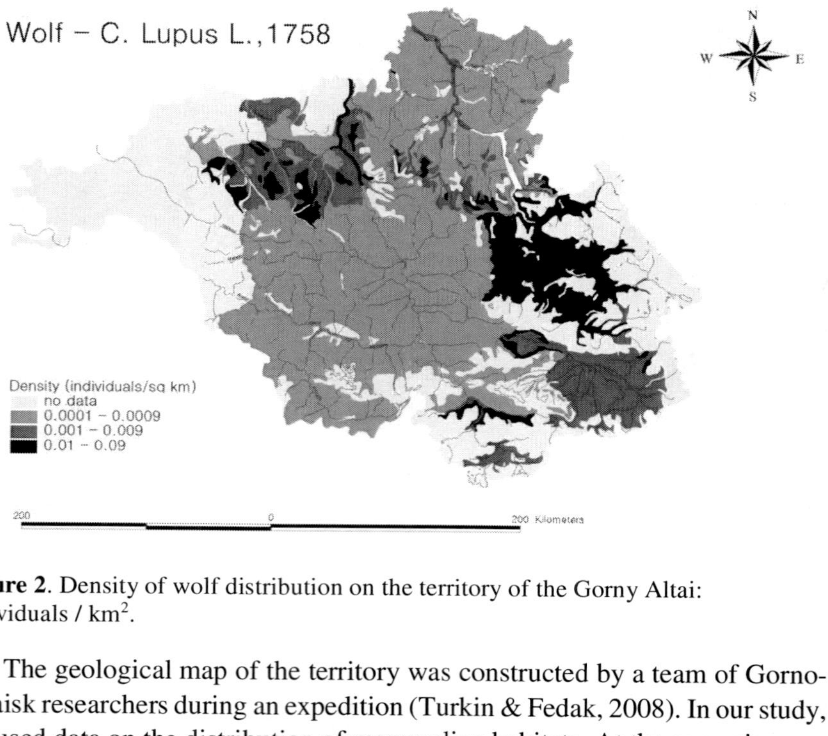


**Figure 2**. Density of wolf distribution on the territory of the Gorny Altai: individuals / km$^2$.

The geological map of the territory was constructed by a team of Gorno-Altaisk researchers during an expedition (Turkin & Fedak, 2008). In our study, we used data on the distribution of mammalian habitats. At the same time we carried out work on the study of the distribution of habitats of mammals in the Altai Mountains, on the basis of geological formations in this study, the number of mammals, which was obtained from the Malkov survey (Figure 2). References and data on the calculation method are given in the Appendix.

Because of the properties of the obtained data (nonlinear, small sample size) nonparametric statistics were used, i.e., Kendall's Rank Correlation (Kendall, 1970).

The following software were used: ArcView 3.2 Geographic Information System, Statistica 6.0 and OpenOffice Calc.

The work was carried out in the following stages:

1. Data on mammal density were collected
2. The geological and landscape maps of Altai Krai (1978) were digitized with ArcView 3.2.
3. Geological and landscape maps were overlaid to form a single layer in ArcView 3.2.
4. Mila Utilities 2.0 (ArcView 3.2) was used to produce a "landscape complex" for each object on the map which was a composite measure of both landscape and geological features (incl soil type, climate, microhabitat)
5. The percentage of the territory of each landscape complex occupied of the total area occupied by a particular rock type was calculated (Table 1).
6. Geological formations which occupied only a small area in the parental landscape (<10km) were filtered out, resulting in a loss of data of less than 3%
7. The table with the data on the density of mammals in landscape complexes and the table containing the data on composition of the landscapes according to their geological formation periods were combined. Kendall's correlation coefficient was calculated between the distribution of geological formations across the landscape complexes and the number of mammals in the landscape complexes using Kendall's correlation.
8. The significance of the correlations was assessed.
9. The resulting data were organised in accordance with their importance, singling out the significant amounts ($p < 0.05$) and grouping them in a form of a table to be further analysed.

**Table 1.** An example of the results of the study of South-Eastern Landscape Province

| Relationship of the area of the landscape complex to the area of the geological formation | Area of the landscape complex within the geological formation (sq.km.) | Percentage of total area of the landscape complex |
|---|---|---|
| High elevational tundra landscape complex / Cambrian period | 45 | 31 |
| High elevational tundra landscape complex / Ordovician period | 100 | 69 |
| Mountain-valley forest landscape complex / Cambrian period | 75 | 49 |
| Mountain-valley forest landscape complex / Ordovician period | 22 | 15 |
| Mountain-valley forest landscape complex / Devonian period | 55 | 36 |

## Results

Kendall's correlation coefficient revealed a number of significant negative correlations between geological composition and mammal density. Furthermore, differences in the extent of the correlations were found between magmatic, metamorphic and sedimentary formations. The highest number of significant correlations occur with intrusive rocks. Taking into account that in different landscape provinces there are mountainous rocks of similar types, we grouped the landscape according to type of rock in order to study the distribution of the types of formation that cause the maximum number of correlations with population numbers - the intrusive rocks – granitoids, diorites, gabbro; the metamorphic rocks – slates, gneisses and the sedimentary rocks – limestone, sandstone (Table 2). The study of the mountainous rocks found in different landscape provinces and grouped according to their composition allows us to conclude that there is a possibility for the influence of the formation-related characteristics of the rocks upon mammal distribution

At the same time the correspondence with the area, occupied by these types of formations, shows that they comprise comparatively insignificant percentage of the area of the landscape provinces (Tables 2 and 3).

**Table 2.** The number of significant correlation coefficients ($p < 0.05$) with the number of population of mammals with regard to the types of the rock

| Landscape provinces | Intrusive rocks | | | Igneous-sedimentary rocks | Metamorphic rocks | | | Sedimentary rocks | |
|---|---|---|---|---|---|---|---|---|---|
| | Granites | Diorites | Gabbro | | A | B | C | D | E |
| South-Eastern | 86 | 6 | 1 | 13 | 5 | 5 | 13 | 44 | 3 |
| Eastern | 17 | 14 | 0 | 41 | 0 | 3 | 17 | 44 | 10 |
| Central | 89 | 17 | 21 | 48 | 15 | 2 | 8 | 24 | 5 |
| North-Eastern | 78 | 29 | 25 | 37 | 13 | 11 | 16 | 14 | 3 |
| Northern | 17 | 3 | 0 | 17 | 3 | 3 | 8 | 47 | 3 |
| North-Western | 36 | 8 | 4 | 11 | 0 | 6 | 0 | 38 | 12 |

A – slates; B – gneisses; C – limestone; D – sandstones, aleurolits; E – sands, bench gravel.

**Table 3.** The area of the mountainous rocks at different landscapes provinces

| Provinces | Total area of province, $km^2$ | Intrusive rocks | | | Igneous-sedimentary rocks, $km^2$ | Metamorphic rocks, $km^2$ | | Sedimentary rocks, $km^2$ | | |
|---|---|---|---|---|---|---|---|---|---|---|
| | | Granites, $km^2$ | Diorites, $km^2$ | Gabbro, $km^2$ | | A | B | C | D | E |
| South-Eastern | 14976 | 2770 | 49 | 16 | 2280 | 48 | 1684 | 69 | 4189 | 3871 |
| Eastern | 14239 | 2033 | 91 | 15 | 405 | 0 | 7171 | 138 | 4305 | 81 |
| Central | 37345 | 3726 | 469 | 305 | 7790 | 725 | 5655 | 644 | 17148 | 983 |
| North-Eastern | 18840 | 1874 | 600 | 1107 | 4684 | 2772 | 1641 | 452 | 5244 | 466 |
| Northern | 13210 | 1764 | 301 | 47 | 857 | 3511 | 51 | 470 | 5543 | 666 |
| North-Western | 16210 | 2702 | 1523 | 53 | 4182 | 0 | 1348 | 0 | 6277 | 123 |

A – slates; B – gneisses; C – limestone; D – sandstones, aleurolits; E – sands, bench gravel.

To reflect the degree of correlation between the area of the geologic formation and the number of mammals we chose a geological formation (granodiorites and biotite melanogenic Kubadrinskyi complex) having the largest spread on the territory of 6 landscape provinces (Table 4). It should be noted that this geological formation has no distribution in the territory of the Northern, North-Western and Central Altai.

As a result of our investigation we composed a map of the degree of the influence of geological formations on mammal density (Figure 4). It is worth noting that the degree of influence is understood as a quantity of significant correlations between the number of population of mammals at different landscapes and the size of the area of every definite formation at the same landscapes. The higher the quantity of the relevant correlations, the higher the influence of the studies formation. Our map reveals the number of coefficients of correlations with regard to every geological formation.

**Table 4.** Example of correlation outcomes τ Kendel (95% confidence level) between the number of mammals and the area of Intrusive geological formation

| Geological formation | Mammal species | Correlation coefficient | | | | | |
|---|---|---|---|---|---|---|---|
| | | East Altai | Northern Altai | Northern - Eastern Altai | North-Western Altai | Central Altai | South East Altai |
| Granodiorites and biotite melanogranites kubadrinskyi complex $\gamma D_{1-2}kb$ | *Alticola macrotis* | 0.64 | | 0.623 | | | 0.424 |
| | *Apodemus agrarius* | | | 0.721 | | | |
| | *Canis lupus* | | | 0.435 | | | |
| | *Capra sibirica* | 0.848 | | | | | 0.477 |
| | *Cervus elaphus* | 0.672 | | | | | 0.388 |
| | *Gulo gulo* | | | 0.725 | | | 0.411 |
| | *Lynx lynx* | | | | | | 0.367 |
| | *Marmota baibacina* | 0.458 | | | | | |
| | *Martes zibellina* | | | | | | |
| | *Microtus agrestis* | | | | | | |
| | *Microtus gregalis* | 0.549 | | | | | 0.431 |
| | *Microtus mongolicus* | | | 0.490 | | | |
| | *Microtus oeconomus* | 0.772 | | | | | 0.733 |
| | *Mustela altaica* | | | | | | |
| | *Myodes glareolus* | | | | | | 0.457 |
| | *Myotis daubentonii* | 0.438 | | 0.481 | | | |
| | *Neomys fodiens* | 0.464 | | | | | |
| | *Ochotona alpina* | 0.864 | | 0.534 | | | 0.537 |
| | *Ovis ammon* | | | 0.515 | | | |
| | *Sorex araneus* | 0.613 | | 0.45 | | | |
| | *Sorex minutissimus* | 0.805 | | 0.42 | | | |
| | *Sorex roboratus Holister* | | | | | | 0.457 |
| | *Sorex tundrensis Merriam* | 0,733 | | 0,633 | | | |
| | *Talpa altaica* | 0.464 | | | | | 0.409 |
| | *Uncia uncia* | 0.742 | | 0.42 | | | |
| | *Ursus arctos* | | | | | | |
| | *Vulpes corsac* | | | 0.459 | | | 0.744 |
| | *Vulpes vulpes* | 0.458 | | | | | |

To show the number of correlations in rocks of different formations, we grouped them by types: Intrusive, effusive, metamorphic, sedimentary rocks.

Then the number of significant correlation coefficients for each type of rock was calculated (Figure 3).

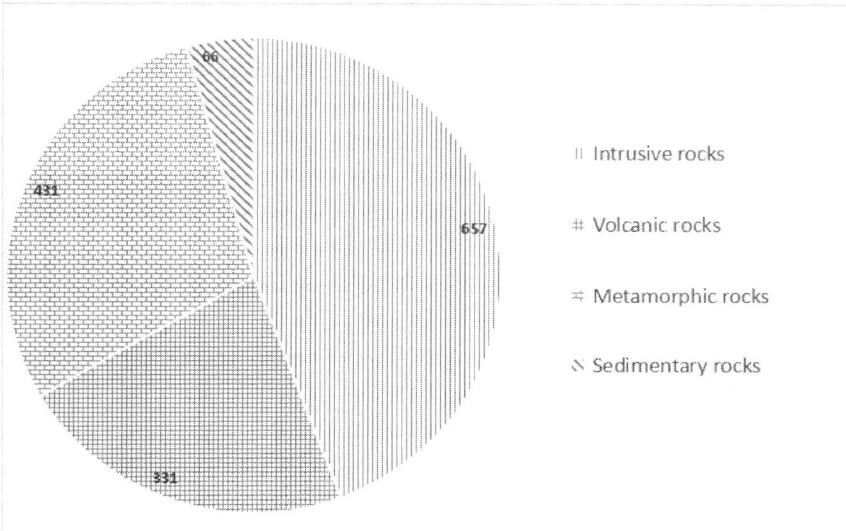


**Figure 3**. Number of significant correlation coefficients by rock types.

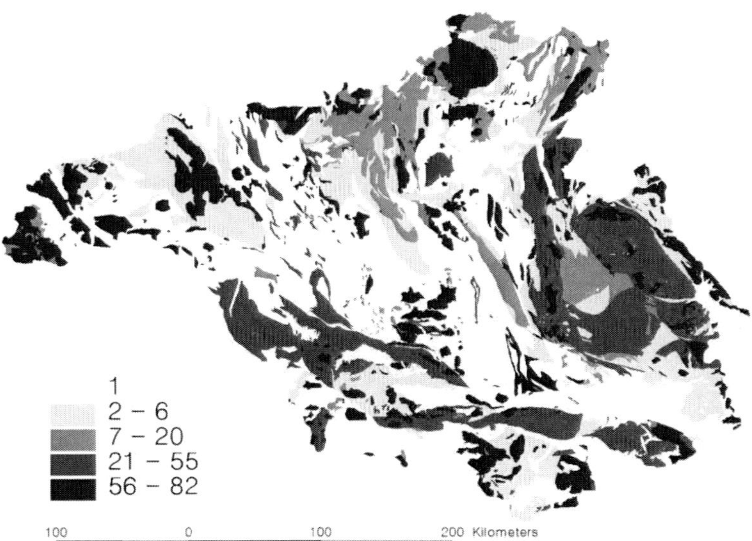


**Figure 4**. The degree of influence of geological formations on mammal numbers - number of significant correlations

## Discussion

Since metamorphic formations make up the largest percentage of the study area, and intrusive rocks only represent 10-20 percent of the area, we conclude that intrusive rocks and the geochemical and geophysical processes connected with them influence the distribution of mammals; these animals tend to avoid areas with high levels of intrusions. It could be argued that the intrusive rocks create some form of relief which particular species of mammals avoid, but our statistical methods have filtered out this type of interaction by holding constant all factors except purely geological ones (rock type) so we have rejected this explanation. Similarly we have rejected climate, soil and vegetation based explanations for the same reason.

The fact that intrusive rocks have the greatest effect on mammal biodiversity and sedimentary and metamorphic rocks have almost no effect on mammal biodiversity suggests that elements such as Fe, Cu, Hg and other metals, may be having an effect, as these elements are usually connected with intrusive arrays. The high level of correlations between mammal distributions and areas containing intrusions (granites, granodiorites, leucogranites, gabbro) suggests the possibility of geochemical and geophysical anomalies and the possible influence of magnetic and electric fields on mammals. The effects of geomagnetism on the migration, homing and alignment of fauna is well documented (Burda et al., 2009; Belova & Acosta-Avalos, 2015; Tian et al., 2015).

The defining feature of the magmatic formation of different stages is its clear geochemical specialization that defines the profile of the formation. That is why the formation-based approach reflects the connection of particular types of hydrothermalites and ores with particular magmatic formations. It is found that the low-depth conditions necessary for endogenous formations define geoecological characteristics of certain territory, whose properties are the contents of lithologic, structural, geochemical, geophysical, hydrothermally changing rocks with regard to its magmatic specialization, and some other characteristics.

Our research shows that the highest number of relevant correlations is found in the territory of the South-Eastern, Eastern and Central Landscape Provinces, where the maximum geodynamic processes for the region are registered (the sum of Neogene-Quarternary deformations is 1500-3000 m) (Atlas, 1978), i.e., it is possible that within the territory of these provinces the additional influence on animal density is caused by geodynamic activity of the territory. We note that at the zones with the active geodynamics the most intensive variations of geophysical fields, the maximal amplitudes of modern movements and the gradients of the stress fields are found. Avoidance responses and changes in distribution of various species around active faults and in earthquake preparation zones have been reported (Subrahmanyam, Sanker Narayan, Srinivasan, 1985; Grant & Halliday 2009; Fidani et al, 2014; Shitov, Borodin et al, 2014; Grant et al., 2015; Pobachenko, Grigoriev et al, 2015; Giannoulopoulou, Evangelou, Karkabounas, Papamarinopoulos, 2018; Jong, Ri, 2022).

Similarly, our hypothesis is indirectly supported by the results of studies on the number of population of rodents in epicentral zone of the Chuya earthquake, which took place within our study area (September 27, 2003; 18:33 M = 7.3). The work reveals considerable influence of geodynamic activity on the behavior and distribution of mammals (Dolgovykh, 2004). Dolgovykh, (2004) reports that:

1. The Siberian ibex (*Capra sibirica*) disappeared from the earthquake area about a month before it took place.
2. Fish left the water in the area of the earthquake. The pattern of this migration was different from the usual seasonal migration patterns. It is possible that such changes may be connected with the changes of the chemical composition of water (V.G. Ushakova, *Pers comm*).
3. According to the evidence of residents of Ongudai Village there were cases when people caught a particular species of freshwater cyprinid fish, (*Oreoleuciscus potanini, O. humilis,)* in an area where it is not usually seen, the mouth of the river Ursul (lat 50,68/ long 86,55 approx). These species usually live at high elevations of the Altai in rivers in the Kosh-Agach and Ulagan Districts. Fishermen reported that they were unable to catch this, usually common, fish between September 26 and October 2. It was also reported that water in the area was discolored with sediment and suspended material.
4. The grey (Altai) marmot (*Marmota baibacina*) that had already entered hibernation at the beginning of the September, woke up in the middle of September, and was noted to be active. After the main shock, it resumed its hibernation.

## Conclusion

Here we have shown how the geochemical, geophysical, geodynamic specialization and other geological characteristics of the underlying rock influences mammal density and distribution. In this process the biological systems (flora and fauna) and underlying geological rocks make up a self-sustaining ecogeological system (ecogeosystem) (Trofimov & Ziling, 2002), so the whole ecosystem is likely to be affected by underlying geology. Our in depth investigation of mammals as an integral part of the natural-territorial complexes improves our understanding of the mechanisms of how internal and interenvironmental correlations operate. In our previous work we have already noted that they are responsible for the information availability and the modification of ecosystems (Karanin, 2004; Karanin, Shitov, 2006).

## Disclaimer

None.

## References

*Atlas of Altai Krai. Vol. 1.* Barnaul – Moscow. (1978). (in Russian).

Begon, M., Townsend, C. R., & Harper, J. L. *Ecology: from individuals to ecosystems* (No. Sirsi) (2006) i9781405111171).

Belova, N. A., & Acosta-Avalos, D. The effect of extremely low frequency alternating magnetic field on the behavior of animals in the presence of the geomagnetic field. *Journal of biophysics* (2015).

Bibikov D. I. methods of counting the number of marmots and experience of its application. Organization and methods of accounting of birds and harmful rodents. Moscow (1963): 192-198. (in Russian).

Brady, K. U., Kruckeberg, A. R., & Bradshaw Jr, H. D. Evolutionary ecology of plant adaptation to serpentine soils. *Annu. Rev. Ecol. Evol. Syst.* (2005):*36* 243-266.

*Brief instructions for the winter route accounting of hunting animals*. Moscow (1986) (in Russian).

Burda, H., Begall, S., Červený, J., Neef, J., & Němec, P. Extremely low-frequency electromagnetic fields disrupt magnetic alignment of ruminants. *Proceedings of the National Academy of Sciences 106*(14) (2009):5708-5713.

Buslov, M., Delvaux D., Zykin V. S., Novikov V. S. Altai Alpine structure: tectonics and evolution // *Active tectonic continental basins: Interaction between structural and sedimentary processes.* Abstracts of the international conference, Gent, Belgium, Apr. 30 – May 2, 1998. − Gent. Belgium: University of Gent, (1998):61–62.

Casazza, G., Barberis, G., & Minuto, L. Ecological characteristics and rarity of endemic plants of the Italian Maritime Alps. *Biological Conservation 123*(3) (2005):361.

Cattaneo-Vietti, R., Albertelli, G., Bavestrello, G., Bianchi, C. N., Cerrano, C., Chiantore, M., ... & Schiaparelli, S. Can rock composition affect sublittoral epibenthic communities?. *Marine Ecology 23* (2002):65-77.

Chasovskikh S. N., Chasovskikh V. A. Account Siberian ibex by direct visual detection // *The Fauna of the Altai-Sayan mountain country*. Gorno-Altaisk (1999):160-163. (in Russian).

Dobretsov N. L., Buslov M. M., Berzin N. A., Simonov V. A. Geology and tectonics of Gorny Altai. Guide-book for post-symposium excursion: The 4-th International Symposium of IGCP Project 283 "*Geodynamic evolution of the Paleoasian Ocean*" // Appendix. – Novosibirsk: UIGGM SD RAS Publ (1993).

Dolgovykh S. V. Biological aspects, noted in Kosh-Agach region at a territory that suffered the influence of the earthquake // The Altai (Chuya) Earthquake: prognoses, characteristics, consequences. Materials of the scientific and practical conference. Ed. by A. V. Shitov. *Gorno-Altaisk* (2004):148-149. (in Russian).

Fidani, C., Freund, F., & Grant, R. Cows come down from the mountains before the (Mw= 6.1) earthquake Colfiorito in September 1997; a single case study. *Animals*, *4*(2), 292-312.

Finkelman R. B., Selinus O., *Centeno J. Environmental Geochemistry and Health*, 29 (2), 2007. – P. 81-82.

Giannoulopoulou L., Evangelou A., Karkabounas S., Papamarinopoulos S. The Effects of Geophysical Anomalies on Biology *Journal of Scientifi c Exploration* 32 2018 3:495–513.

Grant, R. A., & Halliday, T. (2010). Predicting the unpredictable; evidence of pre-seismic anticipatory behaviour in the common toad. *Journal of Zoology*, *281*(4). (2014):263-271.

Grant, R. A., Raulin, J. P., & Freund, F. T. Changes in animal activity prior to a major (M= 7) earthquake in the Peruvian Andes. *Physics and Chemistry of the Earth Parts A/B/C 85* (2015):69-77.

Guidetti, P., & Cattaneo-Vietti, R. Can mineralogical features influence distribution patterns of fish? A case study in shallow Mediterranean rocky reefs. *Journal of the Marine Biological Association of the United Kingdom 82*(6) (2002):1043-1044.

Guidetti, P., Bianchi, C. N., Chiantore, M., Schiaparelli, S., Morri, C., & Cattaneo-Vietti, R. Living on the rocks: substrate mineralogy and the structure of subtidal rocky substrate communities in the Mediterranean Sea. *Marine Ecology Progress Series 274* (2004):57-68.

Jong W, Ri KH. Influence of Geopathic Stress Zone on Heart Rate Variability Parameters. *Int J Clin Exp Physiol*. 2022;9(2):97-100.

Karanin A. V., Shitov A. V. The correlation of geological and geochemical characteristics of landscapes with biodiversity of mammals (with reference to South-Eastern Altai Province) // Newsletter of Tomsk State university. Bulletin of scientific information

*"Relevant questions of geoecology of the Altai and its neighboring regions"* 72 (2006):84-95. (in Russian).

Karanin A. V. *Relationship of mammal biodiversity of the South-Eastern Altai province with geoecological characteristics of landscapes*. Abstract Diss. ... PhD Gorno-Altaisk (2004) (in Russian).

Kendall, M. G. *Rank correlation methods* (4th ed). – London: Griffin (1970).

Kirschvink J. L., Jones D. S., MacFadden B. J. *Magnetite Biomineralization and Magnetoreception in Organisms: a new Biomagnetism* New York: Plenum Press (1985).

Kosinova I. I., Baraboshkina T. A., Kosinov A. E. et al. *Ecological geology of Kursk magnetic anomaly* (KMA). – Voronezh: print office of Voronezh State University (2009) (in Russian).

Kruckeberg AR *Geology and plant life: the effects of land forms and rock types on plants*. University of Washington Press, Seattle, USA (2002).

Kucheruk V. V. New in a technique of the quantitative account of harmful rodents and shrews // The Organization and methods of the account of birds and harmful rodents. Moscow (1963):159-183. (in Russian).

Kucheruk V. V., Tupikova N. V., Evseeva V. S. et al. Experience a critical analysis of methods for quantifying rodents and insectivores with traps / lines // Organization and methods for accounting of birds and harmful rodents. Moscow (1963) (in Russian).

Kuzyakin, A. P. Landscaped areas of the USSR as a basis for zoogeographical survey // proceedings of the conference on questions of zoogeography of the land. Alma-Ata (1960):75-76. (in Russian).

Malkov Yu. P., Belikov V. I. *Mammals of the Altai Republic and Altai Krai.* Gorno-Altaisk (1995) (in Russian).

Malkov, Y. P., Shitov A. V. *Teriological and geographycal Atlas of South-Eastern Altai.* Gorno-Altaisk: GASU (2003) (in Russian).

Malkov, Yu. P. *The Bats Of Altai.* Gorno-Altaisk (1993) (in Russian).

Nikolaev A. V. The study of geophysical processes in the geological sphere at the stage of preparation and development of large-scale seismic catastrophe. Designing of physical bases of methods of man-made influence on local areas in the earth's crust to partially reduce tectonic pressure and decrease of catastrophic earthquakes // Natural environment of Russia: adaptive processes in conditions of changing climate and development of atomic energy: Scientific results obtained in 2012-2014 when the Program No. 4 of fundamental research of the Panel of Russian Academy of Science / Ed. and comp. by A. L. Sobisevich. – Moscow: IPHZ RAN (2014):284-286. (in Russian).

Osipov, V. I. Geoecology as an interdisciplinary science about ecological problems of geospheres // *Geoecology* (1993) 1:4–17. (in Russian).

Persinger, M. A. Geopsychology and geopsychopathology: Mental processes and disorders associated with geochemical and geophysical factors. *Experientia* (1987) *43*:92–104.

Pobachenko S. V., Grigoriev P. E., Sokolov M. V., Shitov A. V. The influence of natural magnetic field inhomogeneity areas of active geological fault on the dynamics of functional state of human brain. Proc. SPIE 9680, 21st International Symposium

Atmospheric and Ocean Optics: Atmospheric Physics, 968050 (November 19, 2015); doi: 10.1117/12.2205284.

Rapant S., Cvečkova V. The potential impact of geological environment on health status of residents of the Slovak Republic Environ. *Geochem. Health*. (2014) 36:543–561.

Ravkin, Y. S., Shvetsov, Y. G., Malkov N. P. Malkov, Yu. p. Density, biomass and diversity of the summer terrestrial vertebrate communities of the Altai // the Model of sustainable socio-economic development of the Altai Republic and the countries of the Altai-Sayan region. Doc. to the Intern. SIMP. *Gorno-Altaisk* (1997):20-36. (in Russian).

Samoilova, G. S. Types of landscapes of mountains of southern Siberia. Moscow: Moscow state University, (1973) (in Russian).

Selinus O., Alloway B. J., Centeno J. A., Finkelman R. B., Fuge, R., Lindh, U., Smedley, P. *Essentials of medical geology, impacts of the natural environment on public health*. Amsterdam: Elsevier Academic (2005).

Shitov A. V. Changes in the Effect of Geological Factors on the Health of the Population during Geodynamical Activation / *Izvestiya, Atmospheric and Oceanic Physics* (2012) 48 8:859–870.

Shitov A. V. Health and people living in a seismically active region faults / Man and geosphere. Ed. I. Florinski. – New York: Nova Science Publishers, Inc.(2010):185-213.

Shitov A. V., Borodin A. S., Tuzhilkin D. A., and Apryatkina M. L. Influence of Physical Fields of Active Geological Faults on the Human Cardiovascular System // *Izvestiya, Atmospheric and Oceanic Physics, (2014) 50 7:696–711.*

Stevanović, V., Tan, K., & Iatrou, G. Distribution of the endemic Balkan flora on serpentine I.–obligate serpentine endemics. *Plant Systematics and Evolution*, *242*(1-4) (2003):149-170.

Sviridov N. S. Vodop'yanov B. G. *Records of fur-bearing animals*. Part 2. Irkutsk (1977) (in Russian).

Subrahmanyam, S., Sanker Narayan, P. V., & Srinivasan, T. M. Eff ect of magnetic micropulsations on the biological systems—A bioenvironmental study. *International Journal of Biometerology*, *29*(3) (1985):293–305.

Tian, L. X., Pan, Y. X., Metzner, W., Zhang, J. S., & Zhang, B. F. Bats respond to very weak magnetic fields. *PloS one*, *10*(4), e0123205.(2015).

Trofimov V. T., Ziling D. G. *The Ecological Geology*. Moscow: Geoinformarket (2002) (in Russian).

Turkin Yu. A., Fedak S. I. *Geology and Structure-Substance Complexes of Gorny Altai*. Tomsk: STT (2008). (in Russian).

Vorozhtsov G. A. The error in the calculation of the number of animals in the mountain taiga hunting and fishing farms // Resources of the animal world of Siberia. Sat.nauch. Tr. Novosibirsk: *Science* (1990): 22-25. (in Russian).

Yudin B. S., Galkina L. I., And Potapkina, A. F. *The Mammals Of The Altai-Sayan Mountain Country*. Novosibirsk: *Science* (1979) (in Russian).

## Supplementary Material

**Table S1.** Distribution of mammals in the landscape provinces of the Gorny Altai

| The name of the landscape province | Species of mammals |
|---|---|
| South East | *Sorex araneus, Sorex roboratus Holister, Sorex caecutiens, Sorex minutissimus, Sorex Roborovskogo, Sorex roboratus, Plecotus auritus, Ochotona dauurica, Ochotona pallasi, Ochotona alpina, Lepus timidus, Lepus tolai, Pteromys volans, Spermophilus undulatus, Marmota baibacina, Allactaga major, Allactaga saltator Eversmann, Dipus sagitta, Apodemus peninsulae, Cricetulus barabensis, Phodopus sungorus, Alticola Strelzowi, Myodes rufocanus, Myodes glareolus, Myodes rutilus, Arvicola amphibius, Microtus socialis, Microtus oeconomus, Microtus arvalis, Microtus mongolicus Radde, Canis Lupus, Vulpes vulpes, Vulpes corsac, Ursus arctos, Martes zibellina, Gulo gulo, Mustela lutreola, Lutra lutra, Mustela erminea, Mustela nivalis, Mustela altaica, Mustela eversmanni, Lynx lynx, Uncia uncia, Felis manul, Sus scrofa, Moschus moschiferus, Cervus elaphus, Capreólus capreólus, Alces alces, Capra capra, Ovis ammon* |
| East | *Talpa altaica, Sorex araneus, Sorex caecutiens, Sorex minutus, Sorex minutissimus, Sorex roborovskogo, Eptesicus nilssonii, Lepus timidus, Ochotona alpina, Pteromys volans, Sciurus vulgaris, Tamias sibiricus, Spermophilus undulatus, Marmota baibacina, Sicista betulina, Apodemus uralensis, Apodemus peninsulae, Mus musculus, Cricetulus barabensis, Plecotus auritus, Myodes rufocanus, Myodes glareolus, Myodes rutilus, Myopus schisticolor, Microtus socialis, Microtus oeconomus, Microtus agrestis, Microtus arvalis, Canis Lupus, Vulpes vulpes, Ursus arctos, Martes zibellina, Gulo gulo, Mustela lutreola, Lutra lutra, Mustela sibirica, Mustela erminea, Mustela nivalis, Mustela altaica, Lynx lynx, Uncia uncia, Sus scrofa, Moschus moschiferus, Capreólus capreólus, Alces alces, Rangifer tarandus, Carpa carpa, Ovis ammon* |
| Central | *Talpa altaica, Sorex araneus, Sorex caecutiens, Sorex minutus, Sorex minutissimus, Crocidura sibirica, Sorex oborovskogo, Plecotus auritus, Eptesicus nilssonii, Myotis mystacinus, Myotis brandtii, Murina leucogaster, Lepus timidus, Pteromys volans, Sciurus vulgaris, Tamias sibiricus, Spermophilus undulatus, Marmota baibacina, Sicista betulina, Sicista subtilis, Ochotona alpina, Apodemus uralensis, Apodemus agrarius, Apodemus peninsulae, Mus musculus, Rattus norvegicus, Phodopus sungorus, Ondatra zibethicus, Alticola Strelzowi, Myodes rufocanus, Myodes glareolus, Myodes rutilus, Arvicola amphibius, Myopus schisticolor, Microtus socialis, Microtus oeconomus, Microtus agrestis, Microtus arvalis, Canis Lupus, Vulpes vulpes, Ursus arctos, Martes zibellina, Gulo gulo, Mustela lutreola, Lutra lutra, Mustela sibirica, Mustela erminea, Mustela nivalis, Mustela altaica, Meles meles, Lynx lynx, Uncia uncia, Sus scrofa, Moschus moschiferus, Cervus elaphus, Capreólus capreólus, Alces alces, Capra capra, Ovis ammon* |

| The name of the landscape province | Species of mammals |
|---|---|
| North-West | *Erinaceus europaeus, Talpa altaica, Sorex araneus, Sorex roboratus Holister, Sorex caecutiens, Sorex minutus, Crocidura sibirica, Sorex Roborovskogo, Plecotus auritus, Myotis ikonnikovi, Eptesicus nilssonii, Murina leucogaster, Sciurus vulgaris, Tamias sibiricus, Spermophilus undulatus, Marmota baibacina, Sicista betulina, Sicista napaea, Lepus europaeus, Ochotona alpina, Ochotona pusilla, Apodemus uralensis, Apodemus agrarius, Apodemus peninsulae, Apodemus agrarius, Micromys minutus, Ellobius talpinus, Cricetulus barabensis, Cricetus cricetus, Myosplax myosplax, Alticola macrotis, Alticola Strelzowi, Myodes rufocanus, Myodes rutilus, Arvicola amphibius, Myopus schisticolor, Microtus gregalis, Microtus oeconomus, Microtus agrestis, Microtus arvalis, Martes zibellina, Mustela lutreola, Lutra lutra, Mustela erminea, Mustela nivalis, Mustela altaica, Mustela putorius, Meles meles, Sus scrofa, Moschus moschiferus, Cervus elaphus, Capreólus capreólus, Alces alces* |
| North | *Erinaceus europaeus, Talpa altaica, Sorex araneus, Sorex roboratus Holister, Sorex caecutiens, Sorex minutus, Sorex minutissimus, Crocidura sibirica, Sorex Roborovskogo, Myotis mystacinus, Myotis dasycneme, Plecotus auritus, Plecotus auritus, Eptesicus nilssonii, Murina leucogaster, Lepus timidus, Lepus europaeus, Ochotona alpina, Sicista subtilis, Pteromys volans, Sciurus vulgaris, Tamias sibiricus, Spermophilus undulatus, Marmota baibacina, Sicista betulina, Apodemus uralensis, Apodemus agrarius, Apodemus peninsulae, Micromys minutus, Rattus norvegicus, Myosplax myosplax, Ondatra zibethicus, Myodes rufocanus, Myodes glareolus, Myodes rutilus, Arvicola amphibius, Myopus schisticolor, Microtus gregalis, Microtus oeconomus, Microtus agrestis, Microtus arvalis, Canis Lupus, Vulpes vulpes, Ursus arctos, Martes zibellina, Gulo gulo, Mustela lutreola, otter, Mustela sibirica, Mustela nivalis, Mustela putorius, Meles meles, Lynx lynx, Sus scrofa, Moschus moschiferus, Cervus elaphus, Capreólus capreólus, Alces alces* |
| North-East | *Erinaceus europaeus, Talpa altaica, Sorex araneus, Sorex roboratus Holister, Sorex caecutiens, Sorex minutus, Sorex minutissimus, Crocidura sibirica, Sorex Roborovskogo, Myotis mystacinus, Myotis brandtii, Myotis ikonnikovi, Myotis frater, Myotis dasycneme, Plecotus auritus, Plecotus auritus, Eptesicus nilssonii, Murina leucogaster, Lepus timidus, Lepus europaeus, Ochotona alpina, Pteromys volans, Sciurus vulgaris, Marmota baibacina, Sicista betulina, Apodemus uralensis, Apodemus agrarius, Apodemus peninsulae, Mus musculus, Micromys minutus, Cricetus cricetus, Rattus norvegicus, Myosplax myosplax, Ondatra zibethicus, Myodes rufocanus, Myodes rutilus, Myopus schisticolor, Arvicola amphibius, Microtus gregalis, Microtus oeconomus, Microtus agrestis, Microtus arvalis, Canis Lupus, Vulpes vulpes, Ursus arctos, Martes zibellina, Gulo gulo, Mustela lutreola, Lutra lutra, Mustela sibirica, Mustela nivalis, Mustela altaica, Meles meles, Lynx lynx, Sus scrofa, Moschus moschiferus, Cervus elaphus, Capreólus capreólus, Alces alces* |

The total volume of material was too small mammals (insectivores and rodents) 194373 masher-days (d/s), 16602 cylinder-days (C/s). Most of the Gorny Altai was surveyed by commercial species (Malkov, Shitov, 2003).

Small mammals include most insectivores and rodents. The research was carried out according to standard methods usingcylinders, cones, juice squeezers of hero-traps (described in Kuzyakin, 1960; Yudin at al, 1979). The species abundance was converted to a unit area of 1 $km^2$ using correction factors. Thus, when catching rodents with high-pressure vessels, the population index calculated at 100 d / s increased by 400 times, cylinders and cones are filled with liquid 145 times, cylinders and cones are not filled with liquid 246.5 times. When catching insectivores with cylinders or cones filled with liquid, the index increases by 115 times, without liquid - by 195 times. Insectivores, caught in the press, were not taken into account, because such catches do not reflect the real number (Kucheruk, 1963; Kucheruk et al., 1963). In the presence of materials for presses and cylinders for the indicator took the average data on the two methods, after their transformation into the area. Hedgehog surveys were carried out in the summer by means of a dog. The capture strip at the account with a dog, in open landscapes, did not exceed 100 m, in the closed wood – twice less. The obtained data were recalculated to the area.

In addition to the above accounting treatment of the cylinders of the fisheries of the Siberian mole, the paper presents records of the traces, underground tunnels and mole (mounds of earth to the surface) with the conversion rates per unit area. Research carried out by I. V. Surkovym (unpublished data), has shown that in an average season on the mainstream caught 4.3 individuals. Therefore. When taking into account the route, it is necessary to count the number of intersections of the main moves. When calculating density per 1 $km^2$. I. V. Zharkov (reference?) took 2 sides of the forest quarter, which corresponds to 2 km of the route along the path or felling. Counting the number of main moves and multiplying by 4.3, he obtained the density index by 1 $km^2$ (Sviridov, Vodopyanov, 1977).

Bats were recorded on the routes on summer evenings during the peak of activity without taking into account the species according to the method proposed by Ravkin (1997). In winter, to clarify the species composition, nearby caves were investigated (Malkov, 1993). When processing the collected material, the proportional ratio of species was taken into account. With respect to the proportion, the abundance of 1 $km^2$ each inhabiting the area has been calculated.

The majority of game mammals were surveyed in winter on the trails. The basis of accounting was adopted “Brief instructions for the winter route accounting of hunting animals” (1986). The density of the species was determined by the formula:

Q = NK/L,

where

Q – indicator of the abundance of individuals per 1 km2;
N - number of traces of the view on the route;
L-route length in km;
K-conversion factor.

**Table S2.** Conversion factor for commercial mammals

| Species of mammals | coefficient |
|---|---|
| Wolf | 0,06*<br>0,45** |
| Korsak | 0,22 |
| Fox | 0,12**<br>0,27* |
| Ermine | 2,25 |
| Kolonok | 0,65 |
| Caress | 0,8 |
| Wolverine | 0,065 |
| Sable | 0,45 |
| Chorus | 0,6 |
| Mink | 1,12 |
| Hare-white | 1,35 |
| Hare-hare | 0,5 |
| Tolai hare | 1,35 |
| ROE deer | 0,75 |
| Moose | 0,72 |
| Maral | 0,75 |
| The musk deer | 1,35 |
| Lynx | 0,27 |
| Squirrel | 5,2 |

* For forest habitats.
** For open habitats.

To record the number of marmots often count the number of animals visually marked on 1 km of the linear route. And also, the number of holes per unit area, using further coefficients of their settlement (the average number of animals living in a hole) (Sviridov, Vodopyanov, 1977). We used a visual method of accounting for gray Groundhog, proposed by Bibikov D.I. (1963), followed by the use of correction factors. Its essence lies in the following. The observer moves along the planned route and counts all the animals encountered on the capture strip in 200 m. The same method was used for badger.

Chipmunk, long-tailed ground squirrels and pikas were surveyed acoustically and visually. When determining the number of birds per unit area, the range of detection of the object was taken into account by S. Ravkin (1997).

The Siberian mountain goat was recorded by direct visual detection in November-December, during the rut period, when the goats form large mixed herds. The data were then recalculated by 1 $km^2$ and extrapolated to the entire landscape, taking into account the cross-country correction factor (Vorozhtsov, 1990; Chasovskikh, 1999).

Chapter 4

# Nutritional Influences on the Population Dynamics of Adjacent Rocky Mountain Bighorn Sheep Populations Showing Differing Trends


**Octavio C. Rosas-Rosas[1,*]**
**and Louis C. Bender[2]**

[1]Colegio de Postgraduados-Campus San Luis Potosí, Iturbide, Salinas, San Luis Potosí, México
[2]Extension Animal Sciences and Natural Resources, New Mexico State University, Las Cruces, New Mexico, USA



## Abstract

Productivity and survival of large mammals are strongly influenced by nutritional condition, yet the effects of condition, diet, or diet quality on population dynamics are often underappreciated. We compared diets and diet quality of two adjacent Rocky Mountain bighorn sheep herds with dissimilar population trends in Rocky Mountain National Park, Colorado. Although diets differed in the use of some forage classes seasonally, most importantly browse, forage species and classes were generally similar (Schoener's index ≥0.82 excluding browse) between herds. Diet quality differed in spring and summer when the declining Mummy Range (MR) bighorn consumed significantly more browse than did the more static Continental Divide (CD) bighorn. Fecal DAPA levels were greater for the CD herd during summer, and indicated that diets of CD bighorn provided ca. 33% more digestible energy than did MR diets. Fecal nitrogen levels were greater for the MR herd during spring and



[*] Corresponding Author's Email: octaviocrr@colpos.mx.



In: Advances in Animal Science and Zoology. Volume 21
Editor: Owen P. Jenkins
ISBN: 979-8-89113-291-7

summer, likely due to precipitation of protein because browse comprised 21–31% of diets of MR bighorn during these seasons. The lower quality diets of MR bighorn were likely due to greater use of browse, which could decrease digestion of cellulose and dietary protein, lowering diet quality during the summer when graminoids dominated the diets of both herds. Because lamb production and survival were similar between herds, nutritional impacts on adult female survival provides a mechanistic explanation for the greater decline of MR bighorn. However, diet quality of both herds was low, consistent with poor lamb recruitment seen in both herds.

**Keywords:** bighorn sheep, DAPA, diets, diet quality, fecal nitrogen, population performance

## Introduction

Nutritional condition is fundamental to productivity and survival of large herbivores and hence their population dynamics, including Rocky Mountain bighorn sheep (*Ovis canadensis canadensis*) (Hanks 1981; National Research Council 1985, 2007; Cook 1990; Blanchard et al. 2003; Beecham et al. 2007; Douhard et al. 2018; Bender 2020). While other factors can affect population dynamics of bighorn, including disease, disturbance, and predation, among others (Beecham et al. 2007), vulnerability to most causes of mortality is directly affected by nutritional condition (Hanks 1981; Festa-Bianchet 1988; Cook 1990; Beecham et al. 2007; Bender 2018, 2020). In turn, nutritional condition is primarily a result of diet quality (National Research Council 1985, 2007); consequently, declines in diet quality can ultimately be responsible for decreased population performance regardless of proximate influences. Despite this, the underlying drivers of species survival and productivity, of which nutrition is predominate, continue to be poorly addressed in studies and models of population performance, which continue to emphasize proximate causes of mortality (Krausman and Shackleton 2000).

Moreover, diet quality is most critical in late spring and summer when forages must provide for the nutritional demands of reproductive females (National Research Council 1985, 2007; Cook 1990; Beecham et al. 2007; Bender 2020); if diet quality is poor, bighorn may exhibit lower productivity (i.e., production and survival of lambs), and productivity accounts for the majority of annual variation in population rate of increase (Gaillard et al. 2001). The energetic costs of lactation can also significantly decrease nutritional condition and survival of females (Piasecke and Bender 2011;

Bender et al. 2013; Bender and Hoenes 2017), and population trend is most sensitive to adult female survival (Gaillard et al. 2000).

Historically, numbers of bighorn in Rocky Mountain National Park (RMNP), Colorado, USA, were below objectives and static or declining (McClintock and White 2007). Two herds, the Mummy Range (MR) herd and the Continental Divide (CD) herd, displayed different trends, with the MR herd showing greater relative declines than the more static CD herd (McClintock 2004; McClintock and White 2007). Although located adjacent to each other, these herds occupied mostly different ranges with minimal intermixing (Baumann 1978; Stevens and Goodson 1993). While many factors are popularly hypothesized to affect the population dynamics of bighorn in RMNP (Beecham et al. 2007), most of these are strongly affected by nutritional condition. To illustrate the potential role of nutrition on the population performance of bighorn in RMNP, we assessed whether relative population trends of MR and CD bighorn could be related to differences in diets in RMNP. Our objectives included: (1) determine seasonal diets; (2) index the quality of seasonal diets; and (3) identifying any differences in diet quality or composition between herds that may influence population trend.

## Study Area

Rocky Mountain National Park is a 1,080 km$^2$ bioreserve located in north-central Colorado, USA. Topography consists of high mountainous peaks interspersed with small subalpine meadows, lakes, streams, glaciers, and tundra at higher elevations. Elevations range from 2400–4345 m. The continental divide bisects RMNP, creating different climatic patterns to the east and west. Precipitation averages 35.1 cm in Estes Park on the eastern side of RMNP, and 50.8 cm in Grand Lake west of the continental divide; 75% falls from April to September. Mean daily high temperatures range from 3.6°C in January to 25.7°C in July in Estes Park. Lower slopes and valleys include mixed montane forest interspersed with bunchgrass and sedge-dominated herbaceous meadows. Subalpine spruce (*Picea* spp.)-fir (*Tsuga* spp.) forests predominate at higher elevations, with tundra and bare rock above timberline.

The MR herd ranges in the Mummy Range Mountains, located in north-central RMNP (Figure 1). Elevations range from 3530–4133 m; terrain is characterized by high peaks separated by broad open alpine meadows. Population of the MR herd was estimated at 61–73 during our study, down from ca. 200 in the 1980s (McClintock and White 2007).

The CD herd ranges in a narrow corridor along the continental divide (Figure 1). Elevations range from 2750–3930 m; terrain is characterized by high open tundra interspersed with alpine and montane forest. McClintock and White (2007) estimated a combined population of the Never Summer and CD herds at 305–312 during our study, down from ca. 450–500 1980s.

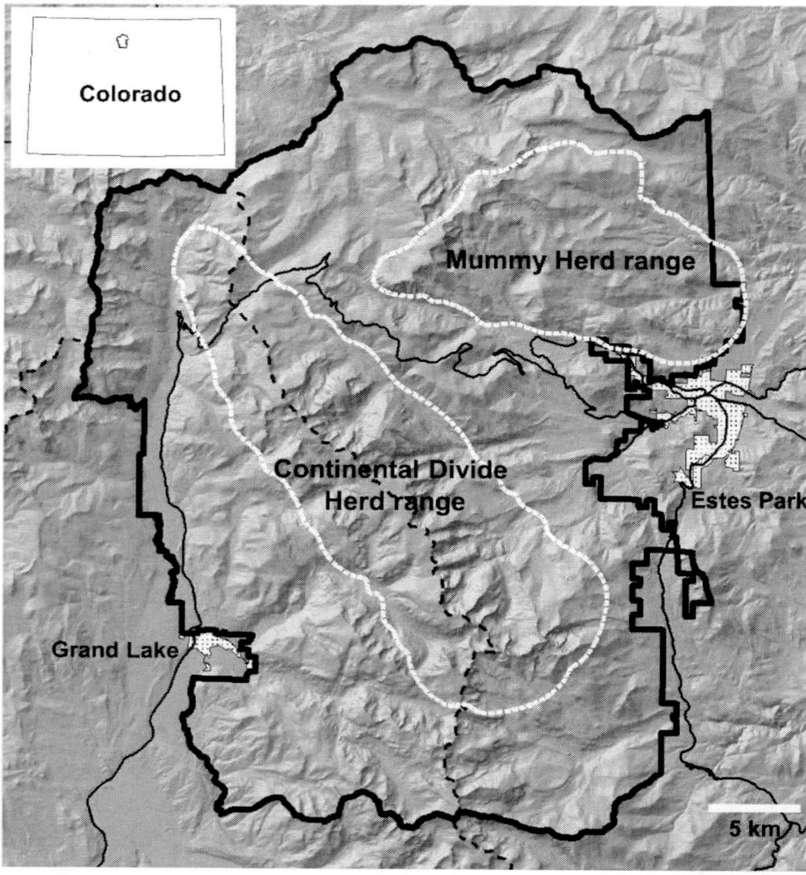


**Figure 1.** General distribution of the Mummy Range herd and the Continental Divide herd in Rocky Mountain National Park (RMNP), north-central Colorado, USA. Insert shows the location of RMNP within the state of Colorado.

# Methods

## Diet Composition and Quality

We collected 253 fresh bighorn fecal samples (153 for MR; 100 for CD), spring 2002–winter 2004. We froze samples as quickly as possible after collection to inhibit bacterial growth and avoid artificially elevated fecal 2,6-diaminopimelic acid (DAPA) levels (Hodgman et al. 1996).

We determined diet composition using fecal microhistological analysis (Holechek et al. 1982b; Wildlife Habitat Laboratory, Washington State University, Pullman, Washington, USA), identifying remains to genus and species when possible. We further categorized remains into 11 classes: grass, sedge (includes rushes), forb, shrub, conifer, moss, fern, seed, lichen, insect, and hair. Because of the low occurrence of the latter 5 classes (< 1% in all seasons), we pooled these into a single class (other).

We used two fecal indices to assess diet quality, fecal nitrogen (FN) and DAPA. FN is correlated with dietary digestibility, dietary protein, and forage intake (Hodgeman et al. 1996; Irwin et al. 1993; Leslie et al. 2008), although consumption of protein complexing compounds such as tannins can artificially inflate FN levels and affect the index (Holechek et al. 1982a; Hobbs 1987; Mould and Robbins 1981; Leslie et al. 2008). DAPA is a component of bacterial cell walls found in the rumen (Van Soest 1994) that passes through the herbivore undigested (Nelson and Davitt 1984). Changes in DAPA reflect changes in rumen bacterial populations that are correlated with intake of digestible energy (DE) and protein (Hodgeman et al. 1996; Nelson et al. 1982; Nelson and Davitt 1984), and are not influenced by the presence of secondary plant metabolites such as tannins (Osborn and Ginnett 2001).

## Data Analysis

We grouped samples into 4 seasons that corresponded with climatic and vegetative patterns and biologically significant seasons for bighorn in RMNP: spring (March–May; corresponding with mid–late gestation), summer (June–August; corresponding with primary lactation), autumn (September–November; corresponding with accrual of fat reserves), and winter

(December–February; the time of lowest forage availability and quality). Exploratory analyses indicated no differences in diet quality indices or diet composition within herds or seasons among years, so we pooled samples across years. We then compared diet quality indices and forage classes among herds and seasons using MANOVA (Morrison 1990), specifically testing the herd × season interaction. If we found differences in diet classes, we conducted further analyses for individual forage species. Last, if we detected seasonal differences in diet quality, we explored relationships among forage classes and diet quality indices within seasons and herds using Pearson's correlations (Zar 1996) and calculated Schoener's similarity index (Schoener 1970) for forage classes and species in diets seasonally between herds.

# Results

## Diet Composition

Proportion of most forage classes in diets differed (herd × season interaction: $F_{7,238} = 14.2$; $P < 0.001$) between herds (Table 1). Grasses comprised 23–66% of seasonal diets and varied seasonally between herds ($F_{7,238} = 6.9$; $P < 0.001$); CD bighorn consumed more grasses during spring ($P = 0.038$), while MR bighorn consumed more grasses during autumn ($P = 0.001$) and winter ($P = 0.004$) (Table 1). Among individual species, only bluegrass (*Poa* spp.) ($F_{7,182} = 3.9$; $P = 0.001$) differed between herds; use was greater ($P < 0.001$) by MR bighorn during autumn ($P \geq 0.222$ for other seasons).

Sedges comprised 18–53% of seasonal diets and varied seasonally between herds ($F_{7,238} = 6.0$; $P < 0.001$) (Table 1). Proportion of sedges/rushes was greater in CD bighorn diets in all seasons ($P \leq 0.046$). Only use of true sedges (*Carex* spp.) varied between herds ($F_{7,233} = 4.7$; $P < 0.001$). CD bighorn consumed more in autumn ($P = 0.046$) and winter ($P < 0.001$); other seasons were similar ($P \geq 0.116$).

**Table 1.** Mean percentages of forage species and forge class totals (SE) of diets of Mummy Range (MR) and Continental Divide (CD) bighorn sheep herds, Rocky Mountain National Park, Colorado, 2002–2004

| Class | Species | Spring | | Summer | | Autumn | | Winter | |
|---|---|---|---|---|---|---|---|---|---|
| | | MR | CD | MR | CD | MR | CD | MR | CD |
| Grass | *Agropyron spp.* | 0.5 | 0.6 | 2.6 | 1.3 | 0.6 | --- | 4.6 | 3.7 |
| | *Agrostis spp.* | 2.1 | 5.4 | 2.0 | 1.7 | 1.4 | 3.5 | 1.9 | 3.6 |
| | *Deschampsia spp.* | 1.5 | 1.8 | 4.1 | 3.1 | 4.9 | 4.1 | 0.5 | 1.2 |
| | *Festuca spp.* | 4.1 | 8.2 | 7.6 | 8.0 | 14.0 | 10.8 | 7.9 | 2.5 |
| | *Phleum spp.* | --- | 0.5 | 0.2 | 0.2 | 4.3 | 0.8 | --- | 1.1 |
| | *Poa spp.* | 6.8 | 9.9 | 8.4 | 10.5 | 28.2 | 2.7 | 8.5 | 6.5 |
| | *Puccinellia spp.* | 4.5 | 0.3 | 1.5 | 0.7 | 5.8 | --- | 24.7 | 0.2 |
| | **Total (SE)** | 23.2 (2.8)A | 31.2 (2.1)B | 34.0 (3.8)A | 32.3 (2.4)A | 65.5 (8.4)A | 35.6 (6.1)B | 53.2 (5.5)A | 33.9 (3.2)B |
| Sedge | *Carex spp.* | 35.3 | 47.3 | 35.0 | 42.4 | 18.5 | 43.6 | 16.2 | 52.3 |
| | *Scirpus spp.* | 0.2 | 0.4 | --- | 0.2 | 1.5 | 0.9 | 1.4 | 0.6 |
| | **Total (SE)** | 37.2 (3.8)A | 49.0 (2.7)B | 36.7 (4.4)A | 45.6 (2.3)B | 20.7 (7.1)A | 45.2 (7.0)B | 17.8 (4.1)A | 53.3 (4.1)B |
| Forb | *Brassica spp.* | 1.1 | --- | --- | --- | --- | --- | 0.2 | --- |
| | *Equisetum spp.* | 0.2 | 0.1 | 3.1 | 0.3 | 1.0 | --- | --- | --- |
| | *Plantago spp.* | --- | 0.7 | 0.1 | 0.4 | --- | 3.3 | --- | 1.0 |
| | *Trifolium spp.* | 0.9 | 1.9 | 0.4 | 2.4 | --- | --- | --- | --- |
| | Unknown forbs | 2.6 | 3.5 | 2.2 | 2.4 | --- | 1.6 | 0.6 | 0.8 |
| | **Total (SE)** | 7.2 (1.4)A | 7.8 (0.8)A | 6.9 (1.6)A | 7.2 (1.0)A | 1.0 (0.6)A | 5.0 (1.9)A | 0.8 (0.3)A | 2.3 (1.1)A |
| Shrub | *Artemisia tridentata* | --- | 0.3 | 0.4 | 1.0 | --- | --- | --- | 0.3 |
| | *Artemisia frigida* | 3.6 | 0.2 | 0.4 | 0.1 | 3.3 | 0.2 | 5.6 | 0.3 |
| | *Artemisia spp.* | 4.2 | 0.3 | 1.2 | 3.0 | 3.6 | 2.8 | 7.0 | 1.8 |
| | *Berberis repens* | --- | --- | --- | --- | --- | 0.6 | 1.7 | --- |
| | *Cercocarpus spp.* | 13.8 | 2.9 | 10.3 | 3.5 | --- | 6.5 | 0.7 | 0.4 |
| | *Ribes spp.* | 1.7 | 0.2 | 3.3 | 0.3 | --- | --- | --- | --- |

| Class | Species | Spring | | Summer | | Autumn | | Winter | |
|---|---|---|---|---|---|---|---|---|---|
| | | MR | CD | MR | CD | MR | CD | MR | CD |
| | *Rubus spp.* | 2.9 | 1.3 | 0.3 | 0.8 | --- | 1.9 | --- | --- |
| | *Salix spp.* | 1.4 | | 1.8 | 0.1 | 1.1 | --- | --- | --- |
| | *Vaccinium spp.* | 0.6 | 2.0 | 0.9 | 0.8 | --- | 1.0 | 0.3 | --- |
| | Unknown shrubs | 1.4 | 1.3 | 1.4 | 1.4 | 1.9 | 0.3 | 0.9 | 0.8 |
| | **Total (SE)** | 31.3 (4.4)A | 9.0 (1.5)B | 21.0 (4.8)A | 12.2 (1.7)B | 9.8 (3.6)A | 13.6 (6.7)A | 16.8 (2.7)A | 6.3 (1.0)A |
| Conifer | *Pseudotsuga menziesii* | 0.4 | 0.8 | 0.3 | 0.5 | 1.4 | --- | 10.2 | 2.5 |
| | **Total (SE)** | 0.5 (0.2)A | 0.9 (0.4)A | 0.3 (0.2)A | 0.7 (0.3)A | 1.4 (0.9) | — | 10.7 (5.8)A | 3.3 (1.2)B |
| Moss | *Aulacomnium* | --- | --- | --- | --- | --- | 0.3 | --- | --- |
| | *Polytrichum* | 0.1 | 1.3 | 0.1 | 0.9 | --- | --- | --- | 0.7 |
| | *Sphagnum* | --- | --- | 0.1 | --- | --- | --- | --- | --- |
| | Unknown mosses | 0.1 | 0.6 | 0.3 | 0.6 | --- | 0.3 | --- | 0.2 |
| | **Total (SE)** | 0.2 (0.1)A | 2.0 (0.7)B | 0.5 (0.2)A | 1.5 (0.4)A | — | 0.6 (0.6) | — | 0.9 (0.5) |

[AB]Means with different letters differ ($P \leq 0.05$) between herds within seasons.

Proportion of shrubs in diets ranged from 9–31% seasonally and varied seasonally between herds ($F_{7,238}$ = 6.2; $P$ < 0.001). Diets of MR bighorn contained more shrubs in spring ($P$ < 0.001) and summer ($P$ = 0.030); other seasons were similar ($P \geq$ 0.128) (Table 1). Only alder-leaf mountain mahogany (*Cercocarpus montanus*; $F_{6,60}$ = 2.4; $P$ = 0.043) varied between herds, being more common in MR bighorn diets during spring ($P$ = 0.050) and summer ($P$ = 0.017; other seasons $P \geq$ 0.952).

Forbs comprised $\leq$7.8% of seasonal diets and varied seasonally within herds ($F_{7,238}$ = 2.8; $P$ = 0.008); no differences were present within seasons between herds ($P$ > 0.291) (Table 1). Both herds consumed more ($P$ < 0.050) forbs in spring and summer; no species differed between herds seasonally ($P \geq$ 0.177).

Conifers comprised $\leq$11% of seasonal diets and use varied seasonally between herds ($F_{7,238}$ = 5.9; $P$ < 0.001) (Table 1). MR bighorn used conifers more during winter ($P$ = 0.001; $P \geq$ 0.645 for all other seasons). No individual species differed seasonally between herds ($P \geq$ 0.622).

Bighorn made limited use of mosses ($\leq$ 2.0% seasonally) (Table 1), but use varied seasonally between herds ($F_{7,238}$ = 2.3; $P$ = 0.025). CD bighorn consumed more mosses during spring ($P$ = 0.002; $P \geq$ 0.100 for all other seasons). No species differed seasonally between herds ($P \geq$ 0.167). Last, all other items pooled did not differ seasonally between herds ($F_{7,238}$ = 1.5; $P$ = 0.067).

## Diet Quality

Diet quality varied seasonally between herds ($F_{7,243}$ = 21.1; $P$ < 0.001), with both FN ($F_{7,243}$ = 10.4; $P$ < 0.001) and DAPA ($F_{7,243}$ = 2.7; $P$ = 0.011) showing significant season × herd interactions (Table 2). FN levels were greater for MR bighorn during spring ($P$ = 0.048) and summer ($P$ = 0.012), but similar for other seasons ($P \geq$ 0.571). DAPA differed only during summer, being significantly greater in CD bighorn ($P$ = 0.028; $P \geq$ 0.808 for other seasons).

FN and DAPA were significantly positively correlated for CD bighorn in spring ($r$ = 0.59; $P$ < 0.001), summer ($r$ = 0.26; $P$ = 0.016), and autumn ($r$ = 0.67; $P$ = 0.012), but not winter ($r$ = 0.17; $P$ = 0.520). FN and DAPA were positively correlated only during spring for MR bighorn ($r$ = 0.51; $P$ <0.001; other seasons $P \geq$ 0.248).

**Table 2.** Mean and SE of seasonal fecal nitrogen (%) and 2,6-Diaminopamelic acid (DAPA; mg/g) in fecal samples from the Mummy Range (MR) and Continental Divide (CD) bighorn sheep herds, Rocky Mountain National Park, Colorado, 2002–2004. Significant differences between herds are in bold ($P \leq 0.050$)

| Season | Fecal nitrogen | | | | DAPA | | | |
|---|---|---|---|---|---|---|---|---|
| | CD | | MR | | CD | | MR | |
| | Mean | SE | Mean | SE | Mean | SE | Mean | SE |
| Spring | **2.660** | **0.128** | **3.109** | **0.183** | 0.485 | 0.020 | 0.475 | 0.026 |
| Summer | **2.623** | **0.101** | **3.184** | **0.147** | **0.545** | **0.017** | **0.487** | **0.023** |
| Autumn | 1.911 | 0.180 | 2.157 | 0.043 | 0.483 | 0.046 | 0.467 | 0.022 |
| Winter | 1.114 | 0.068 | 1.603 | 0.062 | 0.420 | 0.028 | 0.424 | 0.026 |

FN levels were positively correlated to the proportion of shrubs in diets of both herds during both spring and summer (Table 3). FN and DAPA were positively correlated with the proportion of forbs in spring diets of CD bighorn only. No other forage classes were positively related to diet quality indices (Table 3).

**Table 3.** Correlations between the proportion of forage classes identified in fecal samples and the levels of fecal nitrogen (FN) and 2,6-Diaminopamelic acid (DAPA), Mummy Range and Continental Divide bighorn sheep herds, Rocky Mountain National Park, Colorado, 2002–2004. Significant correlations are in bold ($P \leq 0.050$)

| Season | Class | Continental Divide | | | | Mummy Range | | | |
|---|---|---|---|---|---|---|---|---|---|
| | | FN | | DAPA | | FN | | DAPA | |
| | | *r* | *P* | *r* | *P* | *r* | *P* | *r* | *P* |
| Spring | Grasses | <–0.01 | 0.983 | –0.12 | 0.392 | –0.30 | 0.057 | <–0.01 | 0.990 |
| | Sedges | **–0.34** | **0.013** | **–0.32** | **0.020** | **–0.33** | **0.033** | **–0.30** | **0.050** |
| | Forbs | **0.32** | **0.021** | **0.34** | **0.014** | 0.23 | 0.149 | 0.02 | 0.919 |
| | Shrubs | **0.56** | **<0.001** | **0.53** | **<0.001** | **0.42** | **0.005** | 0.26 | 0.100 |
| | Conifers | 0.17 | 0.220 | –0.13 | 0.365 | –0.04 | 0.821 | 0.07 | 0.663 |
| | Mosses | –0.34 | 0.118 | 0.01 | 0.926 | –0.28 | 0.074 | –0.10 | 0.527 |
| Summer | Grasses | –0.06 | 0.665 | –0.07 | 0.575 | –0.24 | 0.201 | –0.01 | 0.601 |
| | Sedges | –0.16 | 0.220 | 0.09 | 0.477 | –0.18 | 0.338 | 0.19 | 0.321 |
| | Forbs | 0.03 | 0.840 | –0.22 | 0.082 | –0.10 | 0.599 | –0.15 | 0.422 |
| | Shrubs | **0.29** | **0.020** | 0.12 | 0.357 | **0.44** | **0.015** | –0.06 | 0.756 |
| | Conifers | 0.08 | 0.554 | <–0.01 | 0.943 | 0.26 | 0.166 | 0.25 | 0.185 |
| | Mosses | **–0.31** | **0.013** | –0.03 | 0.827 | –0.31 | 0.096 | –0.15 | 0.436 |

Diets of MR and CD bighorn were similar (Shoener's index >0.60) in summer and spring for major forage classes (0.89 and 0.78, respectively) and individual forage species (0.80 and 0.69, respectively). Similarity indices increased for both forage classes (0.93 and 0.97, respectively) and individual forage species (0.88 and 0.82, respectively) when browse was excluded from analyses.

## Discussion

MR and CD bighorn differed in diet quality only during summer and spring, when MR bighorn consumed significantly more browse (Table 1). While differences in diet quality reflect foods consumed, they need to be interpreted with respect to season, digestive physiology, and limitations of individual diet quality indices (Leslie et al. 2008). Grasses and sedges were the primary components of bighorn diets, comprising >60% and >78% of spring and summer diets of MR and CD bighorn, respectively (Table 1). Bighorn are primarily grazers adapted to forages with high cellulose content (Geist 1971) because of a large rumeno-reticulum and intestinal volume necessary to efficiently digest cellulose (Hoffman 1988; Krausman et al. 1993; Van Soest 1994); high use of graminoids was thus consistent with their digestive strategy. Similar to other bulk feeders, however, bighorn use a variety of forages, depending on seasonal availability (Wishart 1978; Krausman and Shackleton 2000). While browse is usually a minor component of Rocky Mountain bighorn diets except during autumn and winter–early spring (Wishart 1978), bighorn have included up to 66% browse in diets in Colorado (Todd 1975). However, the digestive efficiency associated with a large rumeno-reticulum can be compromised by consumption of foods high in lignin, tannins, and other phenolics, such as browse (Robbins et al. 1987; Van Soest 1994; Leslie et al. 2008).

These considerations likely affected diet quality and indices of diet quality of MR bighorn during summer and spring in two ways. First, lignin in browse can decrease digestion of cellulose, limiting nutrient gains from a diet composed primarily of graminoids (Robbins 1983; Van Soest 1994); maximizing the digestion of cellulose is the primary advantage of a large rumeno-reticulum (Hoffman 1988; Van Soest 1994). Second, browse can contain high levels of protein-complexing compounds (Mould and Robbins 1981; Robbins 1983; Robbins et al. 1987; Van Soest 1994). These bind with dietary protein causing proteins to pass through the digestive system undigested, elevating FN levels (Robbins 1983; Robbins et al. 1987; Leslie et al. 2008).

Given the high similarity (i.e., Schoener's index $\geq$0.88; also see below) in diets between herds aside from use of browse, the most parsimonious reason for differences in DAPA between herds during summer was compromised digestion of cellulose in MR bighorn because of greater use of browse, and thus decreased nutrient gains from their primary food (graminoids) (Robbins 1983; Robbins et al. 1987; Van Soest 1994). If high in protein-complexing compounds, browse use may also have cost MR bighorn some portion of the protein in their diet due to precipitation (see below; Mould and Robbins 1981; Holechek et al. 1982a; Robbins 1983; Robbins et al. 1987; Van Soest 1994). Unlike FN, DAPA is largely unaffected by use of browse (Osborne and Ginnett 2001). Extrapolating from Figure 1 in Nelson and Davitt (1984), DAPA indicated that DE intake was ca. 4.2 Mcal/day for CD bighorn and ca. 3.0 Mcal/day for MR bighorn during summer (ca. 3.1 Mcal/day for both herds in spring). The DE requirements of a 60-kg domestic sheep are ca. 2.7, 4.4, and 6.6 Mcal/day for maintenance, late gestation, and lactation, respectfully (National Research Council 1985, 2007). While the energetic requirements of domestic sheep may not exactly reflect those of bighorn, summer diets of both herds were likely deficient. Diets of CD bighorn, however, provided >33% more DE.

Summer coincided with late gestation and the peak of lactation, when nutritional requirements of females are greatest (National Research Council 1985, 2007; Cook 1990; Bender 2020). Although both herds were below DE levels required by domestic sheep for lactation, the DE in MR diets in particular may have been too low to support late-term fetuses or neonates without significant mobilization of maternal reserves (Landete-Castillejos et al. 2003). Such mobilization would rapidly decrease condition of females (Landete-Castillejos et al. 2003), and consequently increase their vulnerability to most forms of mortality (Beecham et al. 2007; Bender 2018). Moreover, estimated DE of both herds (again from Figure 1 in Nelson and Davitt 1984) in autumn (2.9–3.1 Mcal/day) and winter (2.3 Mcal/day) was near or below maintenance for domestic sheep (ca. 2.7 Mcal/day), levels which would preclude bighorn from recovering reserves lost during late gestation and lactation. Thus, whether interpreted absolutely or relatively, the lower DAPA in MR bighorn diets during summer provides a mechanistic explanation for the greater decline seen in the MR herd (McClintock and White 2007). Similarly, low DE in diets was consistent with the poor recruitment seen in both herds (0.1–0.2 yearlings/adult female; McClintock 2004).

As noted above, browse use may also have cost MR bighorn some portion of the protein in their diet (Mould and Robbins 1981; Holechek et al. 1982a; Robbins 1983; Robbins et al. 1987; Van Soest 1994). Evidence for this would be elevated FN levels, and diets of MR bighorn were higher in FN than were CD bighorn diets in summer and spring. The utility of FN to reflect dietary nitrogen is compromised when >20–25% of the diet is comprised of forages that contain tannins (Holechek et al. 1982a; Leslie and Starkey 1985). Thus, the greater use of shrubs (21–31% v. 9–12%) by MR bighorn during summer (and spring, although this would have affected FN less; see below) likely inflated FN levels (Mould and Robbins 1981; Holechek et al. 1982a).

Further, if the greater FN levels of MR bighorn actually reflected higher protein intake, DAPA should also have been greater in MR bighorn (Nelson et al. 1982; Hodgeman et al. 1996). However, DAPA was greater for CD bighorn during summer and similar between herds during spring. Moreover, when forages that are low in protein complexing compounds dominate diets, FN and DAPA are positively correlated (Leslie et al. 1989; Hodgeman et al. 1996), and they were significantly positively correlated during spring and summer for CD bighorn. In contrast, when browse constitutes a significant proportion of diets, a positive relationship between FN and DAPA is less likely because of precipitation-inflated FN levels. This was seen with MR bighorn; only in spring were FN and DAPA significantly positively correlated, and levels of tannins and other phenolics in mountain mahogany (the primary browse eaten by MR bighorn) are much lower during spring (Nunez-Hernandez et al. 1991). Collectively, this supports that the higher FN levels of MR bighorn diets during summer reflected precipitation of protein rather than greater protein intake.

Despite the above, FN levels provide insight into diet quality of bighorn in RMNP, at least for CD bighorn. Following Irwin et al. (1993), dietary nitrogen for CD bighorn was 2.08–2.13% (13.0–13.3% crude protein [CP]) during spring and summer, indicating that diets barely met the minimum protein requirements of domestic sheep for gestation and lactation (≥13% CP; Herbert 1973; National Research Council 1985, 2007). This was again consistent with the low lamb recruitment seen in both herds (McClintock 2004). Additionally, autumn and winter diets of both herds contained ca. 8.3–10.0% CP and 6.3–7.0% CP, respectively, which were generally below maintenance levels of domestic sheep (>9%; National Research Council 1985, 2007). If requirements of domestic sheep reflect bighorn needs, these levels would inhibit females from recovering reserves lost during late gestation and lactation, increasing predisposition to all forms of mortality (Beecham et al. 2007; Bender 2018, 2020).

Differences in diet quality reflect diets consumed, probably down to the species level, although small sample sizes for individual species make detecting species-level relationships difficult in RMNP. Grasses, sedges, and shrubs dominated the diets of both herds and thus presumably diet quality. Despite significant differences in use of browse during summer, most (9 of 12) species were present in diets of both herds during summer, and CD and MR bighorn differed only in use of mountain mahogany during summer (and spring). Regardless, Schoener's similarity index indicated that use of browse contributed the most to differences in diet similarity between herds in summer and spring.

Use of shrubs was positively correlated with FN levels during spring and summer in both herds, but was positively correlated with DAPA only in CD bighorn during spring ($P = 0.120$ for MR bighorn in spring; Table 3). Alder-leaf mountain mahogany comprised the majority of browse in diets, and levels of tannin and other phenolics in alder-leaf mountain mahogany in spring are approximately ½ those seen later (Nunez-Hernandez et al. 1991). Consequently, greater use of browse by MR bighorn would have impacted FN and diet quality less during spring as compared to summer. Hence, greater use of browse by MR bighorn during summer most influenced the difference in diet quality between herds.

Use of grasses was similar between herds during spring and summer, as were the species in diets, and proportion of grass in diets was not related to diet quality indices. While CD bighorn consumed more sedge in spring and summer, the species used were again similar between herds. Proportion of sedge in diets was similarly negatively related to both DAPA (–0.32 v. –0.30) and FN (–0.34 v. –0.33) in both herds during spring, and similar relationships for both herds indicated that this relationship had little impact on differences in FN between herds. Use of sedges was also not related to diet quality during summer for either herd.

Despite often high nutritional quality (Wagner and Peek 2006), forbs were limited in diets of both herds (≤7.8%), although they were likely underestimated in diets because forbs are highly digestible and have short retention times (Holechek et al. 2004). The proportion of forbs was similar in spring and summer diets of both herds, and the species were also similar (Table 1). While forbs were positively correlated with DAPA and FN in spring diets of CD bighorn, their influence on diet quality differences between herds was likely minimal, a conclusion supported by the poor quality of spring diets for both herds.

Last, conifers (0.3–0.7%) and mosses (0.5–1.5%) comprised a small proportion of diets of both herds, and only mosses differed between herds. While CD bighorn consumed more moss during summer, the lack of correlation with DAPA and the negative correlation with FN in summer (and spring) suggests that this difference contributed little to the differences in diet quality between herds. Further, mosses were likely overestimated in diets, because moss epidermal tissue often fragments during digestion, inflating presence relative to other foods (Dearden et al. 1975; Leslie et al. 1983).

## Management Implications

Nutritional condition has a predominate effect on survival and productivity of large herbivores, and consequently their population dynamics (e.g., Hanks 1981; Verme and Ullrey 1984; Wakeling and Bender 2003; Cook et al. 2004; National Research Council 2007; Bishop et al. 2009; Bender et al. 2013, 2019; Bender and Rosas-Rosas 2016; Bender and Hoenes 2017; Bender 2020). However, understanding relationships between diet quality indices and diets is complex, complicated by foods consumed, digestive physiology, and seasonal effects on plant chemical composition, among others (Leslie et al. 2008). In RMNP, diet quality of MR bighorn during summer was impacted by their greater use of browse, which could have decreased digestion of cellulose and protein, resulting in their lower diet quality during summer when graminoids dominated the diets of both herds. Summer corresponded with late gestation and lactation, and consequently the ability of females to nourish late-term fetuses and neonates. Low diet quality would impact birth attributes and subsequent survival of lambs (Beecham et al. 2007), unless compensated for by significant mobilization of maternal reserves (Landete-Castillejos et al. 2003). Because productivity was similar between herds (McClintock 2004), the poorer diet quality of MR females likely resulted in greater loss of condition during late-gestation and lactation. Poor condition increases vulnerability to mortality (Beecham et al. 2007; Bender 2018, 2020), providing a mechanistic explanation for the greater relative decline of the MR herd.

Why MR bighorn foraged on browse significantly more during summer (and spring) is unknown. Possibly, forest succession resulting from fire exclusion decreased availability of herbaceous forages (Wakelyn 1987; Keller

2005; Bretfeld et al. 2019). Degraded foraging environments due to fire suppression have low abundance of quality forages necessary to meet the nutritional demands of reproductive females (Arnett 1990; Keller 2005; Bretfeld et al. 2019; Bender 2020). Consequently, increasing the quantity and quality of forage was recommended to increase performance of bighorn populations (Arnett 1990).

In RMNP, underlying nutritional deficiencies thus likely drove population dynamics of bighorn, through well documented impacts on vulnerability to virtually all causes-of-mortality and impacts on production and survival of juveniles. While the majority of studies of population dynamics of large mammals primarily continue to document performance or identify proximate causes of mortality, greater attention to underlying factors that drive survival and productivity is needed to fundamentally understand and predict population performance. Studies and models that do not explicitly address or include the fundamental drivers of population performance will continue to provide limited inferences for conservation and management of large herbivores.

## Acknowledgments

This project was funded by the National Park Service. The New Mexico State University Cooperative Extension Service and Agriculture Experiment Station, and the Colegio de Postgraduados provided additional financial support. We thank B. Keller, B. Hoenes, and A. Hoenes for their contributions to this project. All activities were in accordance with New Mexico State University Institutional Animal Care and Use guidelines.

## References

Arnett, E. B. (1990). *Bighorn sheep habitat selection patterns and response to fire and timber harvest in southcentral Wyoming*. M.S. thesis. Laramie, WY: University of Wyoming.

Baumann, T. G. (1978). *Winter ecology of bighorn sheep in the Mummy Range, Colorado.* M. S. Thesis. Fort Collins, CO: Colorado State University.

Beecham, J. J. Jr., Collins, C. P., and Reynolds, T. D. (2007). *Rocky Mountain bighorn sheep (Ovis canadensis): a technical conservation assessment.* USDA Forest Service, Rocky Mountain Region. Available online: http://www.fs.fed.us/r2/projects/scp/assessments/rockymountainbighornsheep.pdf.

Bender, L. C. (2018). *Understanding predation*. Circular 688. Las Cruces, NM: Cooperative Extension Service, New Mexico State University.

Bender, L. C. (2020). Elk, deer, and pinyon-juniper: needs, what works, and what doesn't. In Malcolm, K., Dykstra, B., Johnson, K., Lightfoot, D., Muldavin, E., and Ramsey, R. (compilers). *Symposium proceedings on piñon-juniper habitats: status and management for wildlife – 2016*, (pp. 34–58). Proceedings RMRS-P-77. Ft. Collins, CO: U.S. Forest Service, Rocky Mountain Research Station.

Bender, L. C., and Rosas-Rosas, O. C. (2016). Compensatory puma predation on adult female mule deer in New Mexico. *Journal of Mammalogy*, 97, 1399–1405.

Bender, L. C., and Hoenes, B. D. (2017). Costs of lactation to body condition and future reproduction of free-ranging mule deer. *Mammalia*, 81, 329–338.

Bender, L. C., Boren, J. C., Halbritter, H., and Cox, S. (2013). Factors influencing survival and productivity of pronghorn in a semiarid grassland-woodland in east-central New Mexico. *Human-Wildlife Interactions*, 7, 313–324.

Bender, L. C., Rosas-Rosas, O. C., Hartsough, M. J., Rodden, C. L., and Morrow, P. C. (2019). Effects of predation on adult pronghorn *Antilocapra americana* Antilocapridae in New Mexico, Southwestern USA. *Mammalia*, 83, 248–254.

Bishop, C. J., and others (2009). Effect of enhanced nutrition on mule deer population rate of change. *Wildlife Monograph*, No. 172.

Blanchard, P., Festa-Bianchet, M., Gaillard, J-M., and Jorgenson, J.T. (2003). A test of long-term fecal nitrogen monitoring to evaluate nutritional status in bighorn sheep. *Journal of Wildlife Management*, 67, 477–484.

Bretfeld, M., Franklin, S. B., and Peet, R. K. (2019). Long-term understory vegetation dynamics of mixed aspen forests in Rocky Mountain National Park, USA. *Journal of Vegetation Science*, 30, 121–133.

Cook, J.G. (1990). *Habitat, nutrition, and population biology of two transplanted bighorn sheep populations in southcentral Wyoming*. Ph.D. dissertation. Laramie, WY: University of Wyoming.

Cook, J. G., and others (2004). Effects of summer-autumn nutrition and parturition date on reproduction and survival of elk. *Wildlife Monographs*, No. 155.

Dearden, B. L., Pegau, R. E., and Hansen, R. M. (1975). Precision of microhistological estimates of ruminant food habits. *Journal of Wildlife Management*, 39, 402–407.

Douhard, M., Guillemente, S., Festa-Bianchet, M., and Pelletier, F. (2018). Drivers and demographic consequences of seasonal mass changes in an alpine ungulate. *Ecology*, 99, 724–734.

Festa-Bianchet, M. (1988). Seasonal range selection in bighorn sheep: conflicts between forage quality, forage quantity, and predator avoidance. *Oecologia*, 75, 580–586

Geist, V. (1971). *Mountain sheep: a study in behavior and evolution*., Chicago, IL: University of Chicago Press.

Gaillard, J.-M., Festa-Bianchet, M., Yoccoz, N. G., Loison, A., and Toigo, C. (2000). Temporal variation in fitness components and population dynamics of large herbivores. *Annual Review of Ecology and Systematics*, 31, 367–393.

Hanks, J. (1981). Characterization of population condition. In Fowler, C. W., and Smith, T. D. (editors). *Dynamics of large mammal populations*, (pp. 47–73). Caldwell, NJ: The Blackburn Press.

Hobbs, N. T. (1987). Fecal indices to dietary quality: a critique. *Journal of Wildlife Management*, 51, 317–320.

Hoffman, R. R. (1988). Anatomy of the digestive tract. In Church, D. C. (editor). *The ruminant animal: digestive physiology and nutrition*, (pp. 14–43). Englewood Cliffs, NJ: Prentice-Hall.

Hodgman, T. P., Davitt, B. B., and Nelson, J. R. (1996). Monitoring mule deer diet quality and intake with fecal indices. *Journal of Range Management*, 49. 215–222.

Holechek, J. L., Vavra, M., and Arthur, D. (1982). Relationships between performance, intake, diet nutritive quality and fecal nutritive quality of cattle on mountain range. *Journal of Range Management*, 35, 741–744.

Holechek, J. L., Vavra, M, and Pieper, R. D. (1982b). Botanical composition determination of range herbivore diets: a review. *Journal of Range Management*, 35, 309–315.

Holechek, J. L, Pieper, R. D., and Herbel, C. H. (2004). *Range management: principles and practices.* Upper Saddle River, NJ: Prentice Hall.

Irwin, L. L., Cook, J. G, McWhirter, D. E., Smith, S. G., Arnett, E. B. (1993). Assessing winter diet quality in bighorn sheep via fecal nitrogen. *Journal of Wildlife Management*, 57, 413–421.

Keller, B. J. (2005). *Behavioral implications of bighorn sheep (Ovis canadensis canadensis) use of the Sheep Lakes mineral site, Rocky Mountain National Park, Colorado.* M. S. thesis. Las Cruces, NM: New Mexico State University.

Krausman, P. R., and Shackleton, D. M. (2000). Bighorn sheep. In Demarais, S., and Krausman, P. R. (editors). *Ecology and management of large mammals in North America*, (pp. 517–544). Upper Saddle River, NJ: Prentice Hall.

Krausman, P. R., Wehausen, J. D., Wallace, M. C., and Etchberger, R. C. (1993). Rumen characteristics of desert races of mountain sheep and desert mule deer. *Southwestern Naturalist*, 38, 172–174.

Landete-Castillejos, T., García, A., Gómez, J. A., Gallego, L. (2003). Lactation under food constraints in Iberian red deer *Cervus elaphus hispanicus*. *Wildlife Biology*, 9, 131–139.

Leslie, D. M., Bowyer, R. T., and Jenks, J. A. (2008). Facts from feces: nitrogen still measures up as a nutritional index for mammalian herbivores. *Journal of Wildlife Management*, 72, 1420–1433.

McClintock, B. T. (2004). *The mark-resight method: an application to bighorn sheep in Colorado and a simulation study evaluating estimators*. M.S. thesis. Fort Collins, CO: Colorado State University.

McClintock, B. T., and White, G. C. (2007). Bighorn sheep abundance following a suspected pneumonia epidemic in Rocky Mountain National Park. *Journal of Wildlife Management*, 71, 183–189.

Morrison, D. F. (1990). *Multivariate statistical methods*. Third edition. New York, NY: McGraw-Hill.

Mould, E. D., and Robbins, C. T. (1981). Nitrogen metabolism in elk. *Journal of Wildlife Management*, 45, 323–334.

National Research Council. (1985). *Nutrient requirements of sheep*. Sixth revised edition. Washington, D.C.: National Academy Press.

National Research Council. (2007). *Nutrient requirements of small ruminants: sheep, goats, cervids, and new world camelids*. Washington, D.C.: National Academy Press.

Nelson, J. R., and Davitt, B.B. (1984). A regional cooperative DAPA research and development program. In Nelson, R. W. (editor). *Proceedings of the 1984 Western States and Provinces Elk Workshop, Alberta, Canada*, (pp. 146–163). Edmonton, Alberta: Alberta Fish and Wildlife.

Nelson, J. R., Koes, R. M., Miller, W. H., and Davitt, B. B. (1982). Big game habitat management on a nutritional basis-a new approach. In Britt, T. L., and Theobald, D. P. (editors). *Proceedings of the 1982 Western States and Provinces Elk Workshop, Flagstaff, Arizona*, (pp. 157–166). Phoenix, AZ: Arizona Game and Fish Department.

Nunez-Hernandez, G., and others (1991). Mountain mahogany and cottonseed meal as supplements for grass hay. *Journal of Range Management*, 44, 497–500.

Osborn, R. G., and Ginnett, T. F. (2001). Fecal nitrogen and 2,6-Diaminopimelic Acid as indices to dietary nitrogen in white-tailed deer. *Wildlife Society Bulletin*, 29, 1131–1139.

Piasecke, J. R., and Bender, L. C. (2011). Patterns in the costs of lactation to body condition of free-ranging Rocky Mountain elk (*Cervus elaphus nelsoni*) populations. In Gupta, V. K., and Verma, A. K. (editors). *Animal diversity, natural history, and conservation.*, Volume 1 (pp. 95–111). New Delhi: Daya.

Robbins, C. T. (1993). *Wildlife feeding and nutrition*. 2nd edition. San Diego, CA: Academic Press.

Robbins, C. T., and others (1987). Role of tannins in defending plants against ruminants: reductions in protein availability. *Ecology*, 68, 98–107.

Schoener, T. W. (1970). Nonsynchronous spatial overlap of lizards in patchy habitats. *Ecology*, 51, 408–418.

Stevens, D. R., and Goodson, N. J. (1993). Assessing effects of removals for transplanting on a high-elevation bighorn sheep population. *Conservation Biology*, 7, 908–915.

Todd, J. W. (1975). Foods of Rocky Mountain bighorn sheep in southern Colorado. *Journal of Wildlife Management*, 39, 108–111.

Van Soest, P. J. (1994). *Nutritional ecology of the ruminant*. 2nd edition. Ithaca, NY: Cornell University-Press.

Verme, L.J., and Ullrey, D.E. (1984). Physiology and nutrition. In Halls, L.K. (editor). *White-tailed deer: Ecology and management*, (pp. 91–118). Harrisburg, PA: Stackpole Books.

Wagner, G. D., and Peek, J. M. (2006). Bighorn sheep diet selection and forage quality in central Idaho. *Northwest Science*, 80, 246–258.

Wakeling, B.F., and Bender, L.C. (2003). Influence of nutrition on mule deer biology and ecology. In deVos, J.C., Jr., Conover, M.R., and Headrick, N.E. (editors). *Mule deer conservation: issues and management strategies*, (pp. 93–116). Logan, UT: Jack H. Berryman Press.

Wakelyn, L. A. (1987). Changing habitat conditions in bighorn sheep ranges in Colorado. *Journal of Wildlife Management*, 51, 904–912.

Wishart, W. (1978). Bighorn sheep. In Schmidt, J. L., and Gilbert, D. L. (editors). *Big game of North America*, (pp. 161–171). Harrisburg, PA: Stackpole Books.

Zar, J. H. (1999). *Biostatistical analysis*. Upper Saddle River, NJ: Prentice Hall.

Chapter 5

# Macrocyclic Lactone-Based Tri-Model Therapy to Treat Humpsore in Cattle under a Humid Tropical Island Ecosystem

**P. Perumal***
**A. K. De**
**and D. Bhattacharya**
Division of Animal Science, ICAR-Central Island Agricultural Research Institute, Port Blair, Andaman and Nicobar Islands, India

## Abstract

Stephanofilariasis or humpsore is a chronic parasitic dermatitis in bovine species and causes severe economic losses to the dairy farmers. Various therapeutic methodologies were tried to treat humpsore in the past; however, they were reported to be partially effective against humpsore. Successful therapeutic methodology was developed to treat humpsore by using FDA-approved macrocyclic lactone (Ivermectin), a piperazine derivative (diethylcarbamazine citrate) and an herbal ointment preparation (Himax). Thirty-six cattle (n = 36 affected) were selected and divided into Gr 1: control (infected without treatment; n = 18) and Gr 2: treatment group (infected with treatment with tri-model therapy including ivermectin, diethylcarbamazine citrate and an herbal ointment, n = 18). In Gr 2, treatment to the ailing animals was given for 45 days. Physiological profiles such as rectal temperature (RT), respiratory rate (RR), pulse rate (PR) and skin temperature (ST), heamatological profiles including differential cell count, biochemical profiles such as total protein (TP), albumin (AL), globulin (GL), creatinine (CR), glucose

* Corresponding Author's Email: perumal.ponraj@icar.gov.in.

In: Advances in Animal Science and Zoology. Volume 21
Editor: Owen P. Jenkins
ISBN: 979-8-89113-291-7

(GLU), total cholesterol (CHO) and urea (URE), liver functional enzymes such as aspartate aminotransferase (AST), alanine aminotransferase (ALT) and alkaline phosphatase (ALP), mineral profiles such as calcium, phosphorous, magnesium, iron, copper and zinc, oxidative stress profiles such as total antioxidant capacity (TAC), superoxide dismutase (SOD) and catalase (CAT) and malondialdehyde (MDA) and hormone such as cortisol were estimated on day 30 of post treatment. Lesion was significantly reduced in day 15 of post-treatment and completely healed on day 45 of post-treatment in Gr 2. Tri-model therapy recorded significant improvement in the surface area of the sore as compared control. Physiological profiles, liver functional enzymes, urea, total white blood cell count including differential cell counts, MDA and cortisol were reduced significantly ($p<0.05$) and blood profiles, biochemical profiles, mineral profiles and antioxidant profiles were increased significantly ($p<0.05$) in tri-model therapy treated animals. Therefore, tri-model therapy has the potential to reduce the liver damage induced by humpsore in dairy animals.

**Keywords**: humpsore, macrocyclic lactone, piperazine derivative, cattle, Andaman and Nicobar Islands

## 1. Introduction

Stephanofilariasis is commonly known as humpsore, caused by the filarial worm *Stephanofilaria assamensis* and transmitted by *Haematobia irritans*, *Musca conducens*, *Musca planiceps* and *Musca autumnalis.* Humpsore is a common chronic skin disorder of cattle endemic to hot humid climate of Indian subcontinent including Andaman and Nicobar Islands [1, 2, 3, 4]. Stephanofilariasis is involved in chronic pruritus, hair follicles damage and loss and causes chronic dermatitis in cattle. Humpsore causes severe economic losses to the dairy farmers as this humpsore induced stress causes loss of milk production, growth rate, fertility rate and loss of market value of the animals due to damaged skin [5]. Humpsore typical lesions occur mostly in and around hump region (92.60%) and rarely in udder, sternal area and other parts of the body [6]. Size of sore differs from a few cm to more than 30 cm [6]. Higher prevalence of Stephanofilariasis is observed in exotic and its crossbred (20.17%) than in zebu cattle (16.14%) [7], mostly prevalent in dark-colored cattle [8] and dry season of the year due to abundance of fly population [2, 3, 4]. Different researchers both in India and abroad attempted to treat the disease with different success rate. Some of the therapeutic methodologies include surgical intervention, use of cauterizing agents [9], external application of petroleum jelly [10], subcutaneous administration of Antimosan with application of 1% gentian violet [11], parenteral administration of 8% Trichlorphon [12], combination of antimony potassium tartrate and 4–8% phenothiazine ointment [13], tobacco ointment alone [13], injectable preparation of levamisole hydrochloride with zinc oxide ointment [2], combination of ivermectin, levamisole and Mastilep ointment [14, 15], ivermectin and Topicure spray [16] and ivermectin and topical application of zinc oxide ointment [17]. Till date, a very effective treatment regime for the disease is not available, and we, here, propose a "tri-model therapy" combining a macrocyclic lactone (ivermectin), a piperazine derivative (DECC), and an herbal preparation (Himax ointment), which was found very effective in treating the disease. Ivermectin, a macrolide anthelmintic, is very effective against microfilaria and has less effect on adult filarial worm, whereas diethylcarbamazine (DECC) is used for prophylaxis against filarial worm in canine [18]. Therefore, in the present study, we have combined these two FDA-approved drugs along with Himax ointment as antibacterial and fly repellent with the hypothesis that they will be effective as chemotherapeutic and chemoprophylactic agents against Stephanofilariasis in cattle. Moreover, the effect of the therapy on liver functional enzymes, biochemical profiles, and mineral profiles alterations was evaluated to judge the efficacy of our proposed tri-model therapy.

## 2. Materials and Methods

### 2.1. Therapeutic Agents

Ivermectin injection (HITEK$^{TM}$) was purchased from Kachhela Medex Pvt. Ltd., Nagpur, India. Diethylcarbamazine citrate (DECC) (Hetrazan$^{TM}$) and Himax$^{TM}$ ointment were procured from Wyeth India, Mumbai, India, and Indian Herbs Overseas, Uttar Pradesh, India, respectively.

### 2.2. Animals and Experimental Design

The present study was conducted at the Cattle Breeding Farm, ICAR-Central Island Agricultural Research Institute, Port Blair, Andaman and Nicobar

Islands, India (11.6060° N, 92.7058° E) from December 2020 to January, 2021 (dry season). The experiment was conducted on crossbred cattle (Indigenous and Jersey). Animals with a single circular or irregular-shaped large/chronic wound (>12 cm diameter) were selected for the current experiment. Humpsore was confirmed by detection of microfilaria in skin scraping of the infected animals. The animals were 4–6 years old with body condition score of 4–5 out of 10. A total of thirty-six affected cattle were chosen and were randomly divided into two groups, namely, Gr 1: control (infected without treatment) and Gr 2: treatment group with tri-model therapy with eighteen animals in each group. Experimental animals were maintained under uniform management with semi-intensive system of rearing where they were allowed for grazing from 7:00 a.m. to 12:00 p.m. Feeding and watering was done as per the farm schedule. *Ad libitum* clean drinking water was available. The details of the groups and treatment protocol areas are follows

- Group 1: Control (infected animals without treatment)
- Group 2: Treatment (infected animals with treatment with tri-model therapy);

Humpsore was cleaned with a mild liquid soap (Lifebuoy liquid soap, Hindustan Unilever Ltd., Mumbai, India) and crusts were removed. Papaya (*Carica papaya*) mist was applied on the wound for about 30 min on first two days. Injection of ivermectin @ 200μg/kg body weight was administered three times by subcutaneous route at 15 days interval (day 0, 15 and 30). Herbal antiseptic and fly-repellent ointment (Himax ointment containing Curcuma longa rhizome extract 0.40 g, Ponggamia pinnata seed oil 4.55 g, Cedrus deodara wood oil, and Gandhaka powder 0.50 g per 100 g of ointment) was applied daily on the wound for 45 days. Diethylcarbamazine citrate 6 mg/kg body weight was administered orally daily for 45 days.

## 2.3. Hematological Analysis

Approximately 10 mL of jugular blood sample was collected from each experimental animal in 15 mL sterile polypropylene centrifuge tube containing heparin (20 IU/mL of blood) as anticoagulant. Blood analysis was carried out using Veterinary fully automatic hematological analyzer (Prokan, PE-6800) directly after the samples were received by the research laboratory and within 30 min after samples were collected. Heamatological profiles such as total red blood cells (TRBC), haemoglobin (Hb), erythrocyte sedimentation rate (ESR), packed cell volume (PCV), total white blood cell (TWBC), differential leucocyte count (DLC: lymphocytes, monocytes, neutrophils, eosinophils) and platelets were estimated. Remaining blood samples were centrifuged at 1200 × g for 15 min at 4°C. The plasma samples were separated rapidly, labelled and preserved at -20°C for further analysis of biochemical, liver functional enzymes and mineral profiles.

## 2.4. Biochemical Profiles

Biochemical indices such as total protein, albumin, globulin, albumin/globulin ratio, creatinine, glucose, total cholesterol, urea, liver functional enzymes such as aspartate aminotransferase, alanine aminotransferase, alkaline phosphatase and mineral profiles such as calcium, phosphorous, magnesium, iron, copper and zinc were estimated with commercially available diagnostic kits (ERBA Mannheim, Germany) using biochemistry analyzer (Automated Clinical Chemistry Analyser EM200, ERBA Diagnostics Mannheim GmbH, Germany).

## 2.5. Oxidative Stress and Hormone Profiles

Total antioxidant capacity (709001), superoxide dismutase (706002) and catalase (707002) in the blood plasma were estimated by using Cayman's Catalase assay kit (Cayman Chemical Company, Ann Arbor, MI, USA) as per the manufacturer's guidelines. Malondialdehyde in plasma was estimated as per the method described by Shah and Walker [19] by using TBA (Thiobarbituric acid)-TCA (Trichloro acetic acid) method. Blood cortisol level was estimated using a commercially available ELISA kit (500360, Cayman Chemical Company, Ann Arbor, MI, USA). All these antioxidants, oxidative free radical and cortisol were estimated in a 96-well clear polypropylene microplate using Alere Microplate Reader (Alere Medical Pvt Ltd, India, AM 2100).

## 2.6. Statistical Analysis

In the present study, in order to determine any possible differences in the observed experimental parameters with respect to between control and treatment groups, student t-test was applied using SAS Software (SAS, Version 9.3.1; SAS Institute, Inc., Cary, NC, 2011). Figures represent the non-transformed data. The mean values are expressed as mean ± SEM. Differences were considered significant if $p < 0.05$.

## 3. Result and Discussion

Tri-model treated cows had significantly ($p<0.05$) lower physiological profiles such as RT (6.04%), RR (14.23%), PR (11.65%) and ST (12.75%) than the untreated control cows (Figure 1). Similarly, the blood profiles such as TRBC (16.91%), Hb (22.02%), ESR (24.72%) and PCV (14.84%) were higher and TWBC (14.56%), neutrophil (22.25%), lymphocyte (11.36%), monocyte (25.83%), eosinophil (54.54%) and platelet (9.84%) were lower significantly ($p<0.05$) in tri-model treated cows than in untreated control cows (Figure 2 and Figure 3). Biochemical profiles such as total protein (24.80%), albumin (28.76%), globulin (21.37%), albumin/globulin ratio (28.35%), glucose (15.89%) and total cholesterol (8.34%) were higher and creatinine (12.54%) and urea (14.28%) were lower significantly in tri-model therapy treated than untreated control cows (Figure 4 and Figure 5). Similarly, liver functional enzymes such as AST (14.76%), ALT (31.21%) and ALP (64.05%) were lower in tri-model treated than untreated cows (Figure 6). Mineral profiles such as calcium (7.11%), phosphorous (14.73%), magnesium (18.62%), iron (8.18%), copper (8.16%) and zinc (7.33%) were significantly ($p<0.05$) higher in tri-model treated than untreated control cows (Figure 7 and Figure 8). Results of the antioxidant and oxidative stress profiles revealed that TAC (12.64%), SOD (18.73%) and CAT (22.31%) were significantly ($p<0.05$) lower and MDA (32.63%) was higher in control humpsore affected animals at day 30 (Figure 9). Similarly, cortisol was significantly decreased in treated than untreated control animals (Figure 9). These blood parameters have shown significant ($p<0.05$) positive correlation with TP, AL, GL, GLU, CHO, mineral and antioxidant profiles whereas significant ($p<0.05$) negative correlation was observed with TWBC, DLC, CR, URE, liver functional enzymes and MDA in tri-model treated cows.

Various treatment regiments were attempted to cure humpsore with varied success rate. However, in our study, tri-model therapy (papaya washing + Diethylcarbamazine tablet + Doramectin injection + Himax ointment) used to treat humpsore, revealed successfully cured within 3 weeks of therapy. This treatment improved the health & wellbeing, milk production and fertility rate in crossbred cattle of ANI.

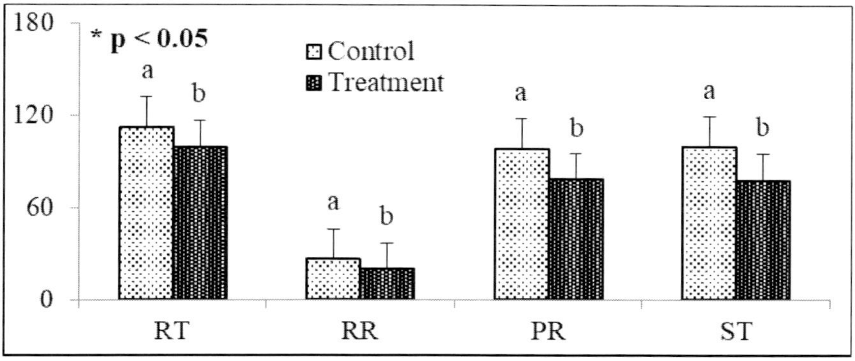


**Figure 1.** Trimodel treatment on physiological profiles in humpsore affected animals (mean ± SEM). Vertical bar on each point represents standard error of mean. Vertical bar with small letters (a, b) indicates significant ($p<0.05$) difference between the control and treatment cows. RT: Rectal Temperature (°F), RR: Respiration rate (bpm), PR: Pulse rate (bpm), ST: Skin Temperature (°F). * indicates $p<0.05$). n= 6 cows for control and treatment cows.

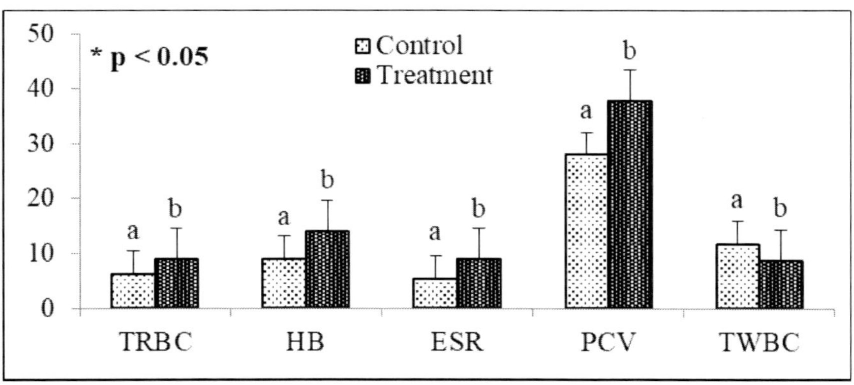


**Figure 2.** Trimodel treatment on heamatological profiles in humpsore affected animals (mean ± SEM). Vertical bar on each point represents standard error of mean. Vertical bar with small letters (a, b) indicates significant ($p<0.05$) difference between the control and treatment cows. TRBC: Total Red Blood Cell ($x10^6/mm^3$), HB: Haemoglobin (g/dl), ESR: Erythrocyte Sedimentation Rate (mm/hr), PCV: Packed Cell Volume (%) and TWBC: total white blood cell ($x10^3/mm^3$). * indicates $p<0.05$). n= 6 cows for control and treatment cows.

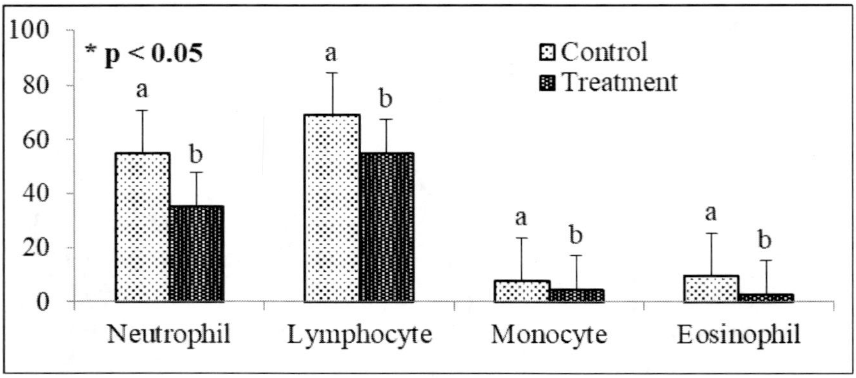


**Figure 3.** Trimodel treatment on Differential cell count (percentage) profiles in humpsore affected animals (mean ± SEM). Vertical bar on each point represents standard error of mean. Vertical bar with small letters (a, b) indicates significant (p<0.05) difference between the control and treatment cows. * indicates p<0.05). n= 6 cows for control and treatment cows.

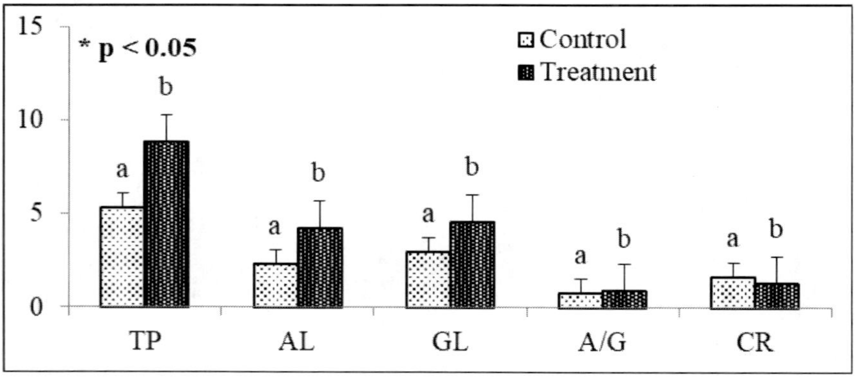


**Figure 4.** Trimodel treatment on blood biochemical profiles in humpsore affected animals (mean ± SEM). Vertical bar on each point represents standard error of mean. Vertical bar with small letters (a, b) indicates significant (p<0.05) difference between the control and treatment cows. TP: Total protein (g/dL), AL: Albumin (g/dL), GL: Globulin (g/dL), A/G: Albumin/Globulin ratio and CR: Creatinine (mg/dL). * indicates p<0.05). n= 6 cows for control and treatment cows.

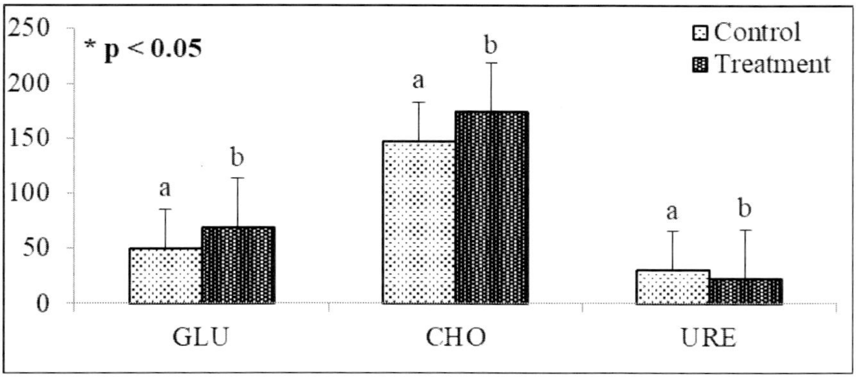


**Figure 5.** Trimodel treatment on blood biochemical profiles in humpsore affected animals (mean ± SEM). Vertical bar on each point represents standard error of mean. Vertical bar with small letters (a, b) indicates significant ($p<0.05$) difference between the control and treatment cows. GLU: Glucose (mg/dL), CHO: Cholesterol (mg/dL) and URE: Urea (mg/dL). * indicates $p<0.05$). n= 6 cows for control and treatment cows.

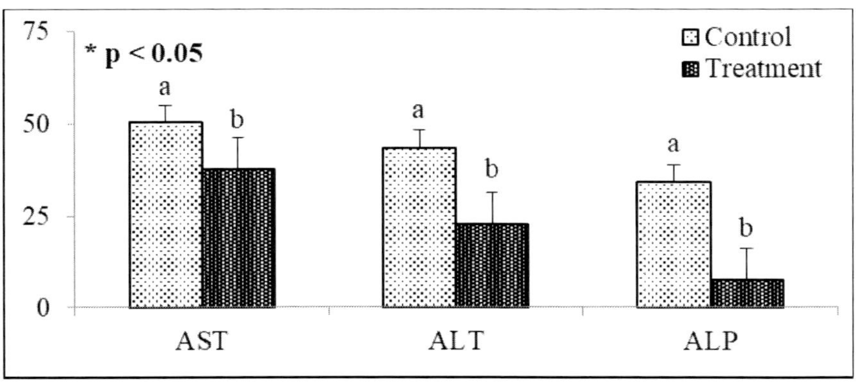


**Figure 6.** Trimodel treatment on liver functional enzymatic profiles in humpsore affected animals (mean ± SEM). Vertical bar on each point represents standard error of mean. Vertical bar with small letters (a, b) indicates significant ($p<0.05$) difference between the control and treatment cows. AST: Aspartate aminotransferase (IU/L), ALT: Alanine aminotransferase (IU/L) and ALP: Alkaline phosphatase (IU/L). * indicates $p<0.05$). n= 6 cows for control and treatment cows.

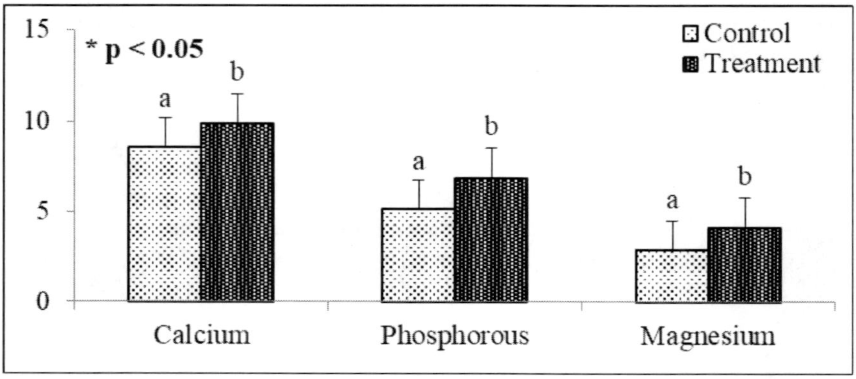


**Figure 7.** Trimodel treatment on blood mineral profiles (mg/dL) in humpsore affected animals (mean ± SEM). Vertical bar on each point represents standard error of mean. Vertical bar with small letters (a, b) indicates significant ($p<0.05$) difference between the control and treatment cows. * indicates $p<0.05$). n= 6 cows for control and treatment cows.

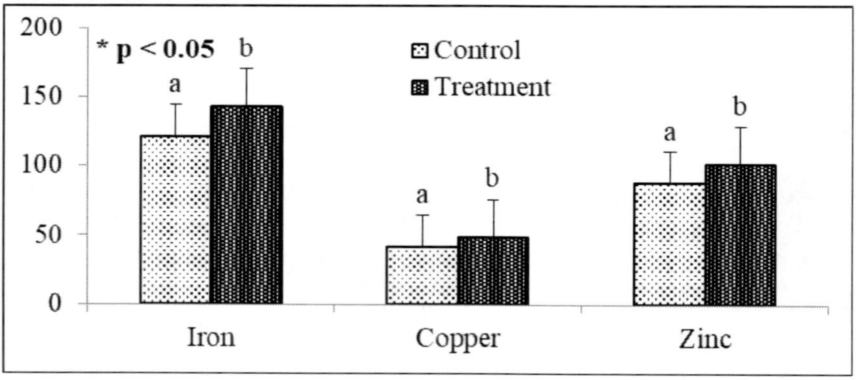


**Figure 8.** Trimodel treatment on blood mineral profiles (µg/dL) in humpsore affected animals (mean ± SEM). Vertical bar on each point represents standard error of mean. Vertical bar with small letters (a, b) indicates significant ($p<0.05$) difference between the control and treatment cows. * indicates $p<0.05$). n= 6 cows for control and treatment cows.

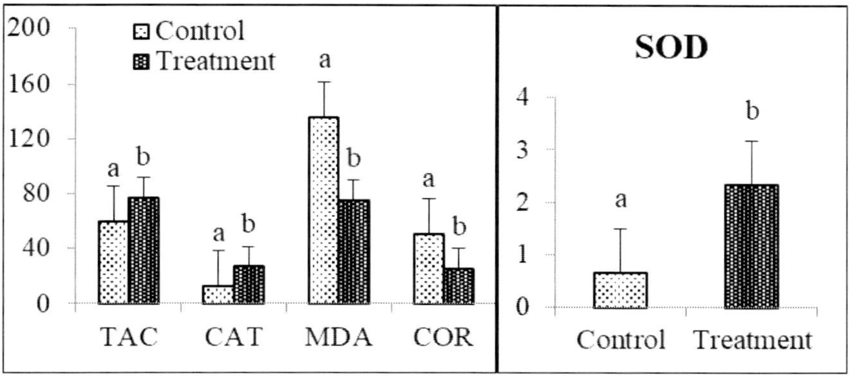


**Figure 9.** Trimodel treatment on liver functional enzymatic profiles in humpsore affected animals (mean ± SEM). Vertical bar on each point represents standard error of mean. Vertical bar with small letters (a, b) indicates significant ($p<0.05$) difference between the control and treatment cows. TAC: Total Antioxidant Capacity (μM/L), CAT: Catalase (nM/min/mL) and MDA: Malondialdehyde (μM/L), COR: Cortisol (ng/mL) and SOD: Superoxide dismutase (U/mL). * indicates $p<0.05$). n= 6 cows for control and treatment cows.

Papaya latex contains enzymes, papain, chymopapain, caricain, glycyl endopeptidase and papaya lipase [20], which has antiviral, antibacterial, antiprotozoal, antifungal, anti-inflammatory and wound healing properties [21]. It desloughs the necrotic tissue and prevents wound infection [22]. Papaya is also used as an ulcer protector [23]. Papain is a cysteine proteinase that digests necrotic tissue by liquefying eschar, thus facilitating the migration of viable cells from the wound edge into the wound cavity [24]. Papain is also reducing the bacterial burden, decreasing exudates and increasing granulation tissue formation [25]. Papain isolated from the latex of unripe papaya pulp is one of the earliest substances used in wound care and chronic skin ulcer therapy as it has anti-bacterial and fibrinolytic properties [26].

DEC is drug of choice to treat the filariasis. It kills both microfilariae and adult worms [27]. Diethylcarbamazine is well absorbed following oral administration. It reaches all parts of the body within 25 minutes after its intake with peak plasma concentrations reached within 1-2 hours. The plasma half-life varies from 6.1 to 8.1 hours. The standard DECC treatment regime is 6 mg/Kg per day over a 10 to 20-day period in filariasis treatment [28]. The elimination half life ranges from 10-12 hours. If the urine is alkalinized, renal excretion of unchanged drug is prevented and the half life of the drug increases. It has an antagonistic action of the metabolic enzymes that metabolize arachidonic acid indicates minimizes the inflammatory condition.

On the other side, DECC involves two processes as in first process, the filarial muscular activity decreases, secondary to hyperpolarization of membranes by the piperazine moiety of DECC and in second process, the microfilarial surface membranes are altered by making them more susceptible to host defenses [29]. Therefore, DECC was administered at 24 h interval to the cattle to treat humpsore for 20-30 days depending upon the degree of the humpsore.

Ivermectin is widely used anti-filarial drug and it is a macrocyclic lactone with broad spectrum activity on filarial parasites. Ivermectin interacts with postsynaptic glutamate-gated chloride channels (GluCl), causes an influx of Cl- ions through the cell membrane by activation of specific ivermectin-sensitive ion channels, resultant muscle hyperpolarization leads paralysis of the microfilarias [30]. Ivermectin acts quickly on the microfilariae and with long-lasting effect, inhibits adult female worms to release additional microfilariae [31]. Dermal microfilarial loads are generally decreased by 78% within two days and by some 98% within two weeks of Ivermectin treatment [32]. Elimination half-life of ivermectin in cattle is 2.0, 3.7, and 8.3 days with aqueous, aqueous: glycerol-formal (50:50 v/v), and propylene glycol: glycerol-formal (60:40 v/v) formulations, respectively [33] whereas its metabolites may persist for more days. Lowest levels of dermal microfilariae observed after this timeframe, it indicates that ivermectin affected microfilariae all are not killed in the first few days. This is supported by reports that microfilariae migrate into deeper dermal layers, sub-cutaneous fat, connective tissue and lymph nodes following administration of the ivermectin [34]. Ivermectin actually interferes with the ability of microfilariae to evade the immune system, resulting in the host's own immune response being able to overcome the immature worms and so kill those microfilariae [35]. It is well known that ivermectin administered SC has persistent for 10-21 days [36, 37, 38]. Therefore, we administered ivermectin at the interval of 15 days (day 0, 15 and 30) in the study to treat humpsore.

Hydrophilic, lipophilic and enzymatic antioxidants are significantly higher in epidermis than in dermis [39]. Ascorbate, urate and glutathione were more than four times higher, ubiquinol-10 was nine times higher and catalase was seven times higher in epidermis indicated that the epidermis is as a biological barrier to prevent the oxidative damage. Normal bacterial flora in the skin generates the ROS. Mechanical injury, abnormal humidity, immunodeficiency or metabolic disorders can result in increased bacterial infection which could result in increased bacterial ROS production. This could act synergistically with ROS from host phagocyte cells to increase the generation of pro-inflammatory agents causing tissue damage [40]. ROS are centrally involved in all wound healing processes as low concentrations of ROS generation are required to fight against invading microorganisms and cell surviving signaling [41]. In line, antioxidant and anti-inflammatory properties of several antioxidant strategies have proven beneficial to improve non-healing state [42]. Ongoing oxidative stress, associated with lipid peroxidation, protein modification and DNA damage has been shown to impair wound healing processes via increased cell apoptosis and senescence [43]. Clinical studies suggest that non-healing wounds are maintained in highly oxidizing environment, which lead to impaired wound repair. Clinical condition such as tissue hypoxia due to occultation of the worm on the passage of the vessels is typically associated with highly oxidizing environments. Therefore, in the present study, significantly higher level of oxidative radical was observed in the humpsore affected cattle than in normal unaffected animals.

Glutathione/glutathione disulphide ratio suggests the level of the oxidation rate and this oxidation rate is increased gradually throughout the life and at the same time, glutathione antioxidant defense system is reduced as age advanced [44]. Such conditions favour to develop ulcerated wound or delay the wound healing as it is occurred in adult or aged animals. Normally during wound healing process, concentration of tissue enzymatic and non-enzymatic antioxidants is reduced as these antioxidants are needed for healing processes [45]. It is not possible to supply antioxidants due to aging for the process of wound healing leads to increased production of ROS which further impose the animals to vulnerability to oxidative stress [46]. It would appear that a physiologically significant contribution to lipophilic antioxidants of skin layers and surface is derived from sebaceous gland secretion [47]. Similarly, in the present study, result revealed that the unaffected animals have significantly higher level of antioxidants than the unaffected animal groups. In the affected animals, the treatment has enhanced the healing process and improved the level of antioxidants and decreased the oxidative stress profile.

RT is a suggestive marker of stress in domestic livestock species [48]. In caprine and bovine species, similar report that RT was increased as the animal was suffered diseases/stresses [49]. Based on the previous literature in other animal species, it was suggested that the livestock suffered diseases were suffered severe stress as the evidence from the physiological parameters. ST was significantly higher in the disease stressed than healthy animal group.

Haematological profiles are important in the determination of physiological changes occurring in animals [50]. Haematological parameters such as PCV, RBC, MCV, MCHC and MCH are used to evaluate the stress/diseased status of animals [51]. Hb and TWBC are also indicators of adaptation to adverse disease stress conditions [50]. Moreover, Hb values are used to assess the degree of stress and wellbeing of animals. The blood parameters were significantly affected in disease stress induced goat as indicated that these animals were suffered with severe haemoconcentration. The higher value in disease affected animal was due to adaptive mechanism exhibited by the animals. Haematological profiles were significantly reduced in diseased than in healthy animals and this was due to the following reasons: (1) as the nuisance and stress increased, body temperature has also been increased which leads to rise in water intake and reduced feed intake [52] leads to deficiency of essential minerals and protein which are important for haemoglobin synthesis, will not be present in the body of the animals in adequate level leading to reduced erythropoiesis, (2) increased body temperature in diseased stage in turn causes the rostral cooling centre of the hypothalamus stimulates the medial satiety centre, which inhibits the appetite centre, resulting in reduced feed intake [53], (3) at high ambient temperature, peripheral vasodilatation and redistribution of cardiac output are associated with expansion of blood volume and haemodilution [54] and (4) destruction of more number of RBC [55]. In general, the oxygen consumption rate is increased when the animal suffering diseases/stress in summer season. Moreover, the animals are moving long distance in the forest to search the feed especially in summer seasons. This leads to increase the oxygen demand to meet the muscular activity. That may lead to increase the RBC production and heamoglobin concentration in livestock. Further, in the experiment, the animals were allowed to drink water or feed the fodder that leads to haemodilution in summer season leads to decrease of the blood parameters particularly RBC, HB, PCV and ESR [56]. Generally during thermal or disease stress, severe dehydration has been reported in livestock which ultimately leads to increased level of Hb and PCV [57]. Further, severe water deprivation in these livestock species during exposed to summer hot season and disease condition could have aggravated the condition. In addition, PCV and Hb are considered to be the indices of the organic response to disease stress in livestock species [48]. Similarly, PCV value was decreased in diseased condition due to excessive RBC destruction and/or haemodilution [58]. Therefore, significantly reduced erythrocytic indices were observed in diseased animals in the present study. Our study revealed that higher TWBC and neutrophils in diseased than in healthy animals. This could be due to release of corticosteroids due to humpsore disease which in turn increased leukocyte count and accelerated mobilization of mature neutrophils from bone marrow storage pool [59]. It also revealed during disease period, lymphocyte count could also be increased [60]. Livestock might be affected sub-clinical parasitic infection in summer which in turn increased leucocyte profiles [61]. Monocyte count was also increased during disease condition could be associated with increased cortisol secretion [62]. Therefore, TWBC was increased during diseased stage. Similar reports were available in literature that higher TWBC in bovine species [63] during diseased state. Negative energy balance and increased blood cortisol concentration during diseased animal leads to immune-suppression which in turn increased susceptibility towards infectious diseases [64]. Therefore, TWBC production was increased in humpsore disease affected cattle in Andaman and Nicobar Islands.

Animals with ulcerated wound as like humpsore may be catabolic and metabolically compromised [65]. Protein and energy substrates are important in maintenance of immuno system and are as active components of several antioxidants [66]. Under conditions of metabolic stress/humpsore disease stress, intake of these micronutrients is important to maintain or restore the tissue viability. However, in many studies, the patients suffered from the ulcerated wound are deficiency of the antioxidants. In our study also, similar report was observed as deficiency of these antioxidants and enhanced oxidative stress profile leads to humpsore condition in cattle. Therefore, supplementation of these antioxidants either through feed or injectable or implant form may be reduced the incidence of the humpsore along with other treatment protocols.

Healing of the epidermis normally involves migration of keratinocytes from the wound edge which undergo proliferation, differentiation and apoptosis. Histopathological studies revealed that hyper proliferative stem cells and actively proliferating keratinocytes at the ulcerated wound margin [67]. Despite the presence of hyper proliferative keratinocytes, healing of the ulcerated wound was slow, suggesting that the problems actually lie with distorted organization of the wound bed. In these patients, it may have been caused by infection and impaired nutritional supply, which impairs keratinocytes migration. ROS may be implicated in impairment of early stages of healing in the wound bed in ulcerated wounds through inflammatory process, with an imbalance of ROS and antioxidant process. Therefore, in the present study, we observed the humpsore wound healing was quicker and faster in the DEC and Doramectin treated group than in untreated group.

## Conclusion

Concluded that trimodel therapy cured the humpsore lesions by controlled the microfilaria as well as adult filarial worm population, besides, improved the antioxidant level and reduced oxidative stress profiles and cortisol concentration. This therapy considerably reduced the stress and improved the health and wellbeing. Therefore, milk yield and fertility rate has been improved considerably in the trimodel therapy treated animals.

## References

[1] Rai, R. B., Srivastava, N., Jai Sunder, Kundu, A. and Jeyakumar, S. 2010. Stephanofilariasis in bovines: Prevalence, control and eradication in Andaman and Nicobar Islands, India. *Indian Journal of Animal Sciences,* 80(6): 500–505.

[2] Rai, R. B., Ahlawat, S. P. S., Surgriv, S., Nagarajan, V. and Singh, S. 1994. Levamisole hydrochloride: an effective treatment for stephanofilarial dermatitis (humpsore) in cattle. *Tropical Animal Health and Production*, 26(3): 175-176.

[3] Rai, R. B. and Ahlawat, S. P. S. 1995. Therapeutic evaluation of levamisole HCl against Stephanofilaria' dermatitis in cattle in Andamans. *Indian Journal of Animal Sciences,* 65(2): 177-179.

[4] Rai, R. B., Ahlawat, S. P. S. and Senani, S. 1995. Seasonal variability in sore characteristics of stephanofilarial dermatitis in Bay Islands. *Indian Veterinary Journal,* 72: 187–188.

[5] Ibrahim, M. Z. U., Hashim, M. A., Hossain, M. A. and Al-Sultan, I. 2003. Comparative efficacy between surgical intervention, organo-phosphorus and ivermectin against humpsore (stephanofilariasis) in cattle. *Journal of Advanced Biomed. Pathobiol. Research,* 3: 100–109.

[6] Rai, R. B., Ahlawat, S. P. S. and Singh S. 1992. Therapeutic evaluation of the efficacy of diethylcarbamazine citrate against stephanofilarial dermatitis in cattle. *Tropical Agriculture,* 69: 1–4.

[7] Singh, K. S., Mukhopadhyay, S. K., Pal, R. N. and Bhattacharjya, M. K. 2002. Studies on the prevalence of stephanofilarial dermatitis in naturally infected cattle in West Bengal with reference to age, sex, breed, season, periodicity and site of lesion. *Journal of Interacademicia,* 6: 646–650.

[8] Mishra, S. P. 1969. Treatment of ear-sore in buffaloes. *Indian Veterinary Journal*, 46: 632.

[9] Patnaik, B. 1970. Studies on stephanofilariasis in Orissa. V. Treatment and control of hump-sore in cattle due to Stephanofilaria assamensis. *Indian Journal of Animal Sciences*, 40: 167-175.

[10] Ahmed, S. and Ali, M. I. 1973. Treatment of humpsore with Antimosan solution. *Veterinary Medical Review,* 2: 136-142.

[11] Baki, M. A. and Dewan, M. L. 1975. Evaluation of treatment of stephanofilariasis (hump sore) with Naguvon (Bayer) by clinical pathological studies. *Bangladesh Veterinary Journal*, 9: 106.

[12] Dutta, P. K. and Hazarika, R. N. 1976. Chemotherapy of stephanofilarial dermatitis in bovines. *Indian Veterinary Journal*, 53: 221-224.

[13] Choudhury, P. S. and Das M. 2012. Comparative efficacy of certain drugs against stephanofilarial dermatitis in cattle. *Journal of Veterinary Parasitology*, 26(1): 60 - 62.

[14] Phukan, S. C., Das, M. and Barkakoty, M. R. 2005. Hump sore in cattle in Assam. *Journal of Veterinary Parasitology*, 19: 19-22.

[15] Puttalakshmamma, G. C., Surendra Sonule and Sumudra. 2012. Hump sore in bullocks in Bidar district – A clinical report. *Journal of Veterinary Parasitology*, 26(1): 82-84.

[16] Islam, M. N., Azad, L. A., Akther, M., Sen, A., Avi, R. D. T. and Juli, M. S. B. 2018. Dermatopathological study of stephanofilariasis (humpsore) in cattle and its therapeutic approaches. *International Journal of Current Research in Life Sciences*, 7(06): 2314-2319.

[17] Dewan, M. L. 1971. Experimental development of stephanofilariasis in zebu cattle. In *Materialy Nauchnoi Konferentsii Povoprosam, Veterinarii.* Part I. Moscow, USSR, Veterinary Acad. USSR. Pp. 83-84.

[18] Sandhu, H. P. 2016. *Essentials of Veterinary Pharmacology and Therapeutics.* 2nd edn. Kalyani Publishers, Noida, India. pp. 1054–1055, 1061–1062.

[19] Shah, J. K. and Walker's, A. M. 1989. Quantitative determination of MDA. *Biochem. Biophys. Acta,* 11: 207–211.

[20] Azarkan, M., El Moussaoui, A., van Wuytswinkel D, Dehon, G. and Loozeet, Y. 2003. Fractionation and purification of the enzymes stored in the latex of *Carica papaya*. *J Chromatogr B Analyt Technol Biomed Life Sci,* 790(1–2): 229–238.

[21] Vij, T. and Prashar, Y. 2015. A review on medicinal properties of *Carica papaya* Linn. Volume 5. *Asian Pacific Journal of Tropical Diseases,* 5: 1–6.

[22] Nayak, S. B., Pinto Pereira, L. and Maharaj, D. 2007. Wound healing activity of *Carica papaya* L. in experimentally induced diabetic rats. *Indian Journal of Experimental Biology,* 45(8): 739–743.

[23] Emeruwa, A. C. 1982. Antibacterial substance from *Carica papaya* fruit extract. *Journal of Natural Products,* 45(2): 123–127.

[24] Telgenhoff, D., Lam, K., Ramsay, S, Vasquez, V., Villareal, K., Slusarewicz, P., Attar, P. and Shroot, B. 2007. Influence of papain urea copper chlorophyllin on wound matrix remodeling. *Wound Repair Regeneration,* 15(5): 727–735.

[25] Falanga, V. 2002. Wound Bed Preparation and the Role of Enzymes: A Case for Multiple Actions of Therapeutic Agents. *Wounds,* 14: 47–57.

[26] Ezike, A. C., Akah, P. A., Okoli, C. O., Ezeuchenne, N. A. and Ezeugwu, S. 2009. *Carica papaya* (Paw-Paw) unripe fruit may be beneficial in ulcer. *Journal of Medicinal Food,* 12(6): 1268–1273.

[27] Dreyer, G., Addiss, D., Williamson, J. and Noroes, J. 2006. Efficacy of co-administered diethylcarbamazine and albendazole against adult *Wuchereria*

*bancrofti*. *Transactions of the Royal Society of Tropical Medicine and Hygiene*, 12: 1118-1125.

[28] World Health Organisation expert committee on filariasis. 1993. Lymphatic filariasis: diagnosis and pathogenesis. WHO expert committee on filariasis. *Bulletin of World Health Organization,* 71: 135.

[29] Kuhlmann, F. M. and Fleckenstein, J. M. 2017. Antiparasitic Agents. In: *Infectious Diseases* ed: Jonathan Cohen, William G. Powderly and Steven M. Opal, 4th Edition, Elsevier Publisher.

[30] Geary, T. G. and Moreno, Y. 2012. Macrocyclic lactone anthelmintics: spectrum of activity and mechanism of action. *Current Pharmaceutical Biotechnology*, 13(6): 866–872.

[31] Taylor, H. R. and Greene, B. M. 1989. The status of ivermectin in the treatment of onchocerciasis. *American Journal of Tropical Medicine and Hygiene*, 41: 460-466.

[32] Plaisier, A. P., Alley, E. S., Boatin, B. A., Van Oortmarssen, G. J., Remme, H., De Vias, S. J., Bonneux, L. and Habbema, J. D. F. 1995. Irreversible effects of ivermectin on adult parasites in onchocerciasis control programme in West Africa. *Journal of Infectious Diseases*, 172: 204-210.

[33] Lo, P. K. A., Fink, D. W., Williams, J. B. and Blodinger, J. 1985. Pharmacokinetic studies of ivermectin: effects of formulation. *Veterinary Research Communications*, 9: 251–258.

[34] Duke, B. O. L., Zea Flores, G., Castro, J., Cupp, E. W. and Mufioz, B. 1992. Effects of three-month doses of ivermectin on adult Onchocerca wlvulus. *American Journal of Tropical Medicine and Hygiene*, 46: 189-194.

[35] Mackenzie, C. D., Geary, T. G. and Gerlach, J. A. 2003. Possible pathogenic pathways in the adverse clinical events seen following ivermectin administration to onchocerciasis patients. *Filarial Journal*, 2(Suppl 1): 5–14.

[36] Barth, D. 1983. Persistent anthelmintic effect of ivermectin in cattle. *Veterinary Record*, 113: 300.

[37] Bremner, K. C., Berrie, D. A. and Hotson, I. K. 1983. Persistence of the anthelmintic activity of ivermectin in calves. *Veterinary Record,* 113: 569.

[38] Armour, J., Bairden, K., Batty, A. F., Davison, C. C. and Ross, D. B. 1985. Persistent anthelmintic activity of ivermectin in cattle. *Veterinary Record,* 116: 151–153.

[39] Shindo, Y., Witt, E., Han, D., Epstein, W. and Packer, L. 1994. Enzymatic and non-enzymatic antioxidants in epidermis and dermis of human skin. *Journal of Investigative Dermatology,* 102: 122-124.

[40] Ginsburg, I. 2002. The role of bacteriolysis in the pathophysiology of inflammation, infection and post-infectious sequelae. *APMIS,* 110: 753-770.

[41] Dunnill, C., Patton, T., Brennan, J., Barrett, J., Dryden, M., Cooke, J., Leaper, D. and Georgopoulos, N. T. 2017. Reactive oxygen species (ROS) and wound healing: The functional role of ROS and emerging ROS-modulating technologies for augmentation of the healing process. *International Wound Journal*, 14: 89–96.

[42] Kunkemoeller, B. and Kyriakides, T. R. 2017. Redox signaling in diabetic wound healing regulates extracellular matrix deposition. *Antioxidants and Redox Signaling*, 27: 823–838.

[43] Bryan, N., Ahswin, H., Smart, N., Bayon, Y., Wohlert, S. and Hunt, J. A. 2012. Reactive oxygen species (ROS)—A family of fate deciding molecules pivotal in constructive inflammation and wound healing. *European Cells and Materials*, 24: 249–265.

[44] Jones, D. P., Mody, V. C. Jr., Carlson, J. L., Lynn, M. J. and Sternberg, P. Jr. 2002. Redox analysis of human plasma allows separation of pro-oxidant events of aging from decline in antioxidant defences. *Free Radicals Biology and Medicine*, 33: 1290-1300.

[45] Shukla, A., Rasik, A. M. and Patnaik, G. K. 1997. Depletion of reduced glutathione, ascorbic acid, vitamin E and antioxidant deference enzymes in a healing cutaneous wound. *Free Radical Research*, 26: 93-101.

[46] Droge, W. 2002. Free radicals in the physiological control of cell function. *Physiology Review*, 82: 47-95.

[47] Thiele, J. J., Weber, S. U. and Packer, L. 1999. Sebaceous gland secretion is a major physiologic route of vitamin E deliver to skin. *Journal of Investigative Dermatology*, 111: 1006-1010.

[48] Sejian, V., Maurya, V. P. and Naqvi, S. M. K. 2010. Physiological adaptability and growth performance of Malpura ewes subjected to thermal and nutritional stress under semiarid tropical environment. *Tropical Animal Health and Production,* 42: 1763–1770.

[49] Coulon, J. B. and Pradel, P. 1997. Effect of walking on roughage intake and milk yield and composition of Montbeliarde and Tarentaise dairy cows. *Annales de Zootechnie,* 46:139–146.

[50] Kumar, B. and Pachaura, S. P. 2000. Haematological profile of crossbred dairy cattle to monitor herd status at medium elevation in central Himalayas. *Research in Veterinary Science,* 69: 141–145.

[51] Koubkova, M., Knizkova, I., Kunc, P., Hartlova, H., Flusser, J. and Dolezal, O. 2002. Influence of high environmental temperature and evaporative cooling on some physiological, haematological and biochemical parameters in high-yielding dairy cows. *Czech Journal of Animal Science,* 47: 309–318.

[52] Quartermain, A. R. and Broadbent, M. P. 1997. Some patterns of response to climate by Zambian goats. *East African Agricultural and Forestry Journal,* 40: 115-124.

[53] Albright, J. L. and Allison, C. W. 1972. Effects of varying the environment upon performance of dairy cattle. *Journal of Animal Science,* 32: 57- 66.

[54] Olsson, K., Josäter-Hermelin, M., Hossaini-Hilali, J., Hydbring, E., & Dahlborn, K. 1995. Heat stress causes excessive drinking in feed and food-deprived pregnant goats. *Comparative Biochemistry and Physiology-Part A,* 110: 309-317.

[55] Shafferi, I., Roussel, J. D. and Koonce, K. X. 1981. Effects of age, temperature, season and breed on blood characteristics of dairy cattle. *Journal of Dairy Science,* 64(1): 63–68.

[56] Sejian, V., Singh, A. K., Sahoo, A. and Naqvi, S. M. K. 2012. Effect of mineral mixture and antioxidant supplementation on growth, reproductive performance and adaptive capability of Malpura ewes subjected to heat stress. *Journal of Animal Physiology and Animal Nutrition,* 98: 72-83.

[57] McManus, C., Paluda, G. R., Louvandini, H., Gugel, R., Sasaki, L. C. B. and Paiva, S. R. 2009. Heat tolerance in Brazilian sheep: physiological and blood parameters. *Tropical Animal Health and Production,* 41: 95–101.

[58] Shebaita, M. K. and Kamal, T. H. 1975. Lean body mass and body fat change in lactating animals under Sahara climate. *World Review on Animal Production,* 23: 28–57.

[59] Jain, N. C. 1986. *Schalm's Veterinary Haematology.* 4th edn. Lea and Febiger, Philadelphia, USA.

[60] Narayan, U., Singh, D. V. and Sharma, R. J. 2007. Physiological responses in Holstein-Friesian crossbred lactating cows under heat stress ameliorating treatments. *Indian Journal of Animal Research,* 41: 130–133.

[61] Rutkowiak, B. 2001. Laboratory blood tests in prophylaxis of non-infectious diseases of cattle – history or necessity. *Zycie Weterynaryjne,* 76: 196-201.

[62] Jain, N. C. 1993. *Essentials of Veterinary Haematology.* 1st edn. Lea and Febiger, Philadelphia, USA.

[63] Saror, D. and Coles, E. H. 1973. The blood picture of White Fulani (Zebu) and White Fulani/Fresian (crossbreed) dairy cow. *Bulletin of Epizootic Diseases of Africa,* 21: 485–487.

[64] Kehrli, M. T., Nonnecke, B. J. and Roth, J. A. 1989. Alterations in bovine lymphocyte function during the periparturient period. *American Journal of Veterinary Research,* 50: 215–220.

[65] Bliss, M. R. 1998. Pressure injuries: causes and prevention. *Hospital Medicine*, 59: 841-844.

[66] Evans, P. and Halliwell, B. 2001. Micronutrients: oxidant/antioxidant status. *British Journal of Nutrition*, 85(suppl 2): S67-S74.

[67] Galkowska, H., Olszewsk, W. L., Wojewodzka, U., Mijal, J. and Filipiuk, E. 2003. Expression of apoptosis and cell cycle related proteins in epidermis of venous leg and diabetic foot ulcers. *Surgery,* 134: 213-220.

Chapter 6

# Is the Decrease in cAMP Concentration in Oocytes Really a Trigger for Oocyte Maturation Induction? - From the Perspective of Fish Research

**Md. Forhad Hossain and Toshinobu Tokumoto***

Integrated Bioscience Section, Graduate School of Science and Technology, National University Corporation Shizuoka University, Shizuoka, Japan

## Abstract

Since the discovery in *Xenopus laevis* that intracellular cAMP levels decrease upon induction of oocyte maturation, the relationship between oocyte meiotic arrest and intracellular cAMP levels has been studied. In amphibians such as *Xenopus* and fish, the oocyte stops meiotic cell division and grows into a giant cell until it is induced to resume meiosis by maturation-inducing hormone (MIH). MIH is a steroid hormone in these organisms. Surprisingly, experimental results have shown that steroid hormones do not act through the generally known nuclear receptor-mediated genomic reaction but through cell surface receptors in the meiosis induction of oocytes. This highlights a link between decreased intracellular cAMP levels and cell-surface G-protein-coupled MIH receptors. After many years of trying to identify the receptor, it became clear that the cell surface MIH receptor is a receptor named membrane progestin receptor (mPR). Subsequent experiments have suggested that mPR is coupled to an inhibitory G protein (Gi), suggesting the mechanism of mPR-mediated initiation of oocyte maturation

* Corresponding Author's Email: tokumoto.toshinobu@shizuoka.ac.jp.

In: Advances in Animal Science and Zoology. Volume 21
Editor: Owen P. Jenkins
ISBN: 979-8-89113-291-7

induction in amphibians and fish. However, in amphibians, androgen-induced pathways and pathways not mediated by cAMP have been shown to exist. Thus, the involvement of cAMP is still the subject of debate.

This chapter summarizes the history and current status of research on the involvement of changes in intracellular cAMP concentrations in the fish and frog oocyte maturation-inducing pathway.

**Keywords:** oocyte maturation, maturation-inducing hormone, cAMP, progesterone, membrane progesterone receptor, mPR, PAQR

## History of Research on the Relationship between a Decreased Intracellular cAMP Concentration and Oocyte Maturation Induction

To achieve early development in the absence of external energy, the egg halts meiosis to store nutrients and it grows into a giant cell. Once egg development is complete, the action of egg maturation-inducing hormone (MIH) causes the oocyte to complete oocyte maturation and undergo meiotic division prior to fertilization. In vertebrates such as frogs and fish, egg maturation is induced by steroid hormones, especially progesterones. In frogs and fish, ovaries can be removed, and the oocytes can be cultured *in vitro*. This experimental system has been used to investigate how steroid hormones induce oocyte maturation. The first finding reported in *Xenopus laevis* showed a decrease in the cAMP concentration in the oocyte in a short period of time after steroid hormone treatment (Maller et al. 1979). Other reports that have detected 10–60% decreases in cAMP levels from 15 seconds to 6 hours after P4 addition have supported these results (Cicirelli and Smith 1985; Gelerstein et al. 1988; Mulner et al. 1979; Sadler and Maller 1982; Schorderet-Slatkine and Baulieu 1982).

Since the reports in *Xenopus*, studies have shown that rapid decreases in cAMP levels occur in oocytes upon the addition of maturing steroids in fishes (Haider and Chaube 1995; Jalabert and Finet 1986). Subsequently, studies using agents that act on the cAMP-producing enzyme adenylate cyclase have been conducted and revealed that hormonal stimulation of MIH inhibits adenylate cyclase activity. The results suggest that MIH receptors are linked to inhibitory G proteins.

Supporting evidence for the involvement of a decrease in cAMP concentration in oocytes in oocyte maturation induction was obtained by using forskolin, an adenylate cyclase activator. Forskolin, a kind of diterpene from plants that activates adenylate cyclase, inhibits MIH-induced oocyte maturation in medaka (Iwamatsu et al. 1987), yellow perch (DeManno and Goetz 1987a), brook trout (Demanno and Goetz 1987b), and catfish (Haider and Chaube 1995). Furthermore, it was shown in *Xenopus* (Sadler and Maller 1982) and rainbow trout (Yoshikuni and Nagahama 1994) that adenylyl cyclase activity in the oocyte membrane is suppressed by the action of MIH. Cholera toxin (CT), a proteinaceous toxin secreted by the cholera pathogen that is composed of 6 subunits (one $\alpha$ subunit and five $\beta$ subunits), was found to inhibit progesterone-induced egg maturation in *Xenopus* by increasing intracellular cAMP levels (Stith and Maller 1984). In *Xenopus* oocytes, pertussis toxin (PT), a proteinaceous toxin produced by *Bordetella pertussis* that is composed of 6 subunits (no. 1, 2, 3, 5 and 6 subunits and two 4 subunits), was shown to participate in progesterone inhibition of adenylate cyclase via a sensitive G protein (Pellaz and Schorderet-Slatkine 1989). These results suggest that steroid-mediated oocyte maturation induction is initiated by the reduction in cAMP concentration in the oocyte via the G protein. Furthermore, the relationship between egg maturation induction and G protein was established. The molecular characteristics of the G protein response to MIH action have since been analyzed.

The mechanism by which the MIH for fish, 17$\alpha$, 20$\beta$-dihydroxy-4-pregnen-3-one (DHP), acts on receptors on the outer surfaces of rainbow trout oocytes to induce oocyte maturation was investigated in rainbow trout (Yoshikuni and Nagahama 1994), and the types of G proteins interacting with DHP receptors were clarified in medaka fish (Oba et al. 1997). PT catalyzes ADP-ribosylation of a 40 kDa protein in oocyte membranes after vitellogenesis in rainbow trout, and CT catalyzes labeling of several proteins, including a major protein with an apparent molecular weight of 43 kDa. The 40 kDa protein was recognized by antibodies against the $\alpha$ subunit of inhibitory G proteins (Gi), whereas the 43 kDa protein was recognized by antibodies against the $\alpha$ subunit of stimulatory G proteins (Gs). Treatment of the membrane fraction with DHP reduced PT-catalyzed ADP-ribosylation of the 40 kDa protein. In contrast, there was no significant change in CT-catalyzed ribosylation of the 43 kDa protein after exposure to DHP. Specific binding of DHP to membrane fractions was reduced by PT. DHP binding was also inhibited by nonhydrolyzable GTP analogs such as guanosine 5'-O-(3-thiotriphosphate) (GTP gamma S) and guanylylimidodiphosphate (GppNHp), but it was not inhibited by ATP or guanosine 5'-O-(2-thiodiphosphate) (GDP beta S). From these results, it was concluded that PT-reactive Gi is involved in the signal transduction pathway by DHP in rainbow trout oocytes. In medaka, oscillation of the expression of Giα protein on the plasma membrane was found to occur during the induction of oocyte maturation (Oba et al. 1997). Thus, it has been hypothesized that inhibitory G proteins mediate egg maturation induction via MIH action on the basis of studies using frog and fish oocytes.

## Evidence for the Membrane Receptor of MIH and Identification of mPR and Its Coupling with Gi

### Nongenomic Actions of Steroids

Steroid hormones are classified into five types: glucocorticoids and mineralocorticoids produced by the adrenal glands, androgens, estrogens, and progesterones, which are mainly synthesized in the gonads. Analogs of these hormones have been developed as pharmaceuticals and used as anti-inflammatory and reproductive medicines. Steroid hormones act as chemical signaling molecules on a wide variety of target tissues to induce various physiological responses. These include slow genomic responses and fast, acute nongenomic responses. Steroid hormones are associated with cytoplasmic or nuclear receptors. The action of steroid hormones introduced in textbooks is the genomic reaction. Fat-soluble steroid hormones can pass through cell membranes to penetrate cells. Steroid hormones that enter a cell bind to nuclear receptors and enter the nucleus. Genomic DNA (chromosomal DNA) has a special sequence that allows nuclear receptors, including steroid hormone-bound nuclear receptors, to bind. RNA polymerase can then bind at the same site, and mRNA is synthesized using the genomic DNA of the part bound by the nuclear receptor as a template. In this way, in cells that are exposed to a certain steroid hormone, a change is induced in which mRNA is synthesized from a gene with a sequence that allows binding of the corresponding nuclear receptor. Proteins synthesized from mRNA whose expression is specifically induced bring new functions to the cell. This reaction is the genomic reaction that induces gene expression from genomic DNA. The synthesis of mRNA and subsequent protein synthesis are involved in this reaction, so the reaction takes several hours or days.

On the other hand, it is known that the pharmacological action of steroid hormones causes a rapid reaction that cannot be explained by genomic action, as it occurs within a few minutes. Since nongenomic action is a rapid reaction that can be induced in just a few minutes, it is presumed to involve post-translational modifications that cause structural changes in proteins, such as phosphorylation and acetylation. Receptors that mediate nongenomic actions include membrane progestin receptor (mPR), a 7-transmembrane steroid membrane receptor described below, as well as progesterone receptor membrane associated component 1 (PGRMC1), a single-transmembrane receptor that has also been identified (Xu et al. 2011). In addition to these cell membrane receptors, there are also reports that nuclear receptors bound to the inside of the cell membrane act as receptors for nongenomic reactions.

Nongenomic responses have long been thought to be enigmatic, but the structures of membrane receptors have been known for more than 20 years. The discovery of a 7-transmembrane receptor with a unique structure paved the way for the elucidation of the molecular mechanism of nongenomic actions.

## Oocyte Maturation Research That Led to the Discovery of Steroid Membrane Receptors

Spawning in fish and amphibians is ultimately triggered by steroid hormones. The first change during oviposition is resumption of oocyte meiosis, or oocyte maturation. After that, the oocytes that have completed oocyte maturation pass through the follicular cells that have supported their growth up to that point, undergo ovulation and are laid outside the body. Although there are several types of steroid hormones, progesterones and progestins act to induce oocyte maturation. DHP, a kind of progestin, was the first steroid identified in vertebrates as a fish oocyte maturation-inducing hormone (Nagahama and Adachi 1985). In the well-studied *Xenopus*, progesterone is thought to be the egg maturation-inducing hormone (Sadler and Maller 1985). As described above, steroid hormones have hitherto been understood to bind to nuclear receptors and act as transcription factors, that is, to act through genomic actions. However, in 1977, Ishikawa et al. of Shizuoka University proposed that the action of steroid hormones in inducing oocyte maturation is a nongenomic action from the cell membrane surface (Ishikawa et al. 1977). In the case of frog and fish oocytes, oocyte maturation can be experimentally induced *in vitro* using oocytes removed from the ovaries. Since oocyte maturation, which can be regarded as the origin of development, can be experimented upon with a relatively simple operation, many studies have been conducted to elucidate the mechanism. Oocytes stop dividing during meiosis and grow into very large cells over months to years. When the oocyte approaches its maximum diameter (1.2 mm in *Xenopus*), it becomes sensitive to oocyte maturation-inducing hormone. Oocytes at this stage are removed and placed in culture medium; progesterone, an oocyte maturation-inducing hormone, is added to the culture medium; and meiosis is induced by the action of this hormone. Whether meiosis has occurred can be confirmed by checking whether the nucleus (germinal vesicle) in the oocyte has undergone division and disappeared. Using this oocyte maturation induction reaction as a foundation, researchers have investigated the mechanism by adding various agents, such as hormones and enzyme inhibitors, to the culture medium. In addition, for the purpose of examining the reactions in oocytes, oocyte cytoplasm has been collected with a thin glass needle and extracted by centrifugation to examine changes in the components in the cytoplasm. One of the most notable achievements was the discovery of a maturation-promoting factor (MPF) that is activated in the ooplasm after hormonal action (Masui and Clarke 1979; Masui and Markert 1971). Masui et al. of the University of Toronto collected the cytoplasm of oocytes that had undergone oocyte maturation via hormone treatment and injected it into oocytes before hormone treatment. They found that the injected oocytes were activated by the cytoplasm from mature oocytes. This factor that was activated in mature oocytes was named maturation-promoting factor (MPF). Subsequent research has revealed that the same molecule works in somatic cell division, and its presence in various organisms, including yeast, has been confirmed. The discovery of MPF led to the elucidation of its activation mechanism. The intracellular reaction pathway after progesterone action was rapidly elucidated. However, the action site (receptor) of progesterone, which participates in the first reaction, remains under investigation. The first proposal of the presence of progesterone receptors on the cell surface was published in 1977 as a result of a study synthesizing cell-membrane-impermeable steroid hormones. Ishikawa et al. synthesized a hormone-agarose complex that could not pass through cell membranes by chemically bonding steroid hormones to grains of agarose (Ishikawa et al. 1977). Although it was not possible to bind agarose to progesterone itself, chemical crosslinking was possible with deoxycorticosterone, which has a similar structure to progesterone and induces oocyte maturation. For such syntheses, the distance between hormone and agarose was important. The hormone could not bind to the receptor without the proper distance. Oocyte maturation was induced when the synthesized cell-membrane-impermeable steroid was added to the culture solution (Figure 1).

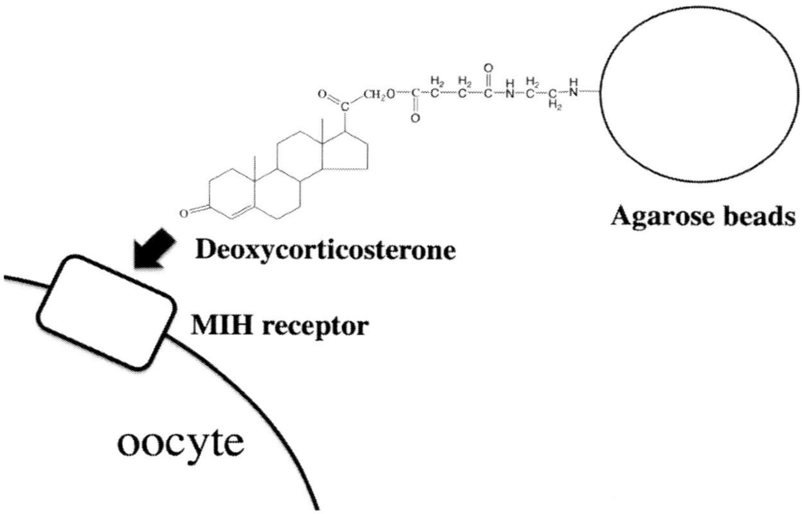


**Figure 1.** Induction of oocyte maturation by agarose bead-coupled deoxycorticosterone. An insoluble steroid derivative prepared by the coupling of deoxycorticosterone with aminoethylated agarose beads is indicated. The receptor for maturation-inducing hormone (MIH) is indicated on the cell membrane of oocytes.

This result indicated that the steroid induced oocyte maturation by binding to the receptor from outside the oocyte; that is, the receptor for oocyte MIH existed on the cell membrane. Experiments also showed that injection of progesterone into oocyte cells did not induce oocyte maturation in frogs (*Rana pipiens*) (Masui and Markert 1971). Slightly later, similar experimental results were published (Godeau et al. 1978). After that, research advanced toward the discovery of the cell membrane receptor for progesterone, but the extraction of the cell membrane protein was generally difficult, and the research progressed slowly.

## Discovery of the Progesterone Membrane Receptor, mPR

Although research was gradually decreasing due to difficulty, two laboratories in Japan and the United States continued research. Twenty-six years after its existence was proposed, screening using monoclonal antibodies against membrane protein fractions finally succeeded in identifying the membrane progestin receptor, mPR. A cDNA sequence for a membrane receptor for oocyte maturation-inducing steroids was reported by Peter Thomas at the

University of Texas in 2003 (Zhu et al. 2003). Antibodies recognizing candidate molecules were isolated via selection of those that inhibited the steroid binding of MIH from a group of monoclonal antibodies prepared against partially purified membrane proteins and immuno-screening using those antibodies.

Based on the deduced amino acid sequence of the candidate gene and subsequent analysis using the gene, it was found that this gene possessed the properties of the oocyte MIH receptor that had been presumed to exist from previous experiments. In other words, the recombinant protein synthesized by introducing the gene into cultured cells had a high binding affinity for progestin. Furthermore, conclusive evidence was obtained that experiments inhibiting protein synthesis from the mRNA of this gene (knockdown experiments with morpholino antisense oligos) inhibited oocyte maturation. These facts strongly suggested that this molecule was the MIH receptor itself that oocyte maturation researchers had been pursuing. This molecule was named membrane progestin receptor (mPR). A gene database search revealed that there are three types of mPR genes that are commonly present in animals ranging from humans to fish, which were classified as subtypes mPRα, mPRβ, and mPRγ (Zhu et al. 2003). It was also revealed that mPR is expressed not only in oocytes but also in various tissues, including the brain and kidneys. After that, the analyses of the genome sequences of various species were further advanced. It was shown that mPRs share homology with the receptors for AdipoQ, a hormone involved in processes regulating the amount of visceral fat, such as the regulation of sugar uptake into cells and the burning of fat (Yamauchi et al. 2003). There are 11 types of homologous genes in the genomes of various organisms, which were classified into a new receptor family, the progestin and AdipoQ receptor (PAQR) family (Tang et al. 2005).

Within this family, three types of mPR molecules, α, β, and γ, correspond to PAQR7, PAQR8, and PAQR5, respectively (Tokumoto et al. 2006; Tokumoto et al. 2012). Subsequent studies have shown that PAQR6 and PAQR9 also have progestin-binding activity, designated mPRδ and mPRε (Pang et al. 2013). At present, there are five types of mPR genes, paqr5-9. Among these, mPRδ and mPRβ are expressed in the hippocampus of the brain, and mPRε is highly expressed in the pituitary gland, suggesting that each subtype has different functions in controlling brain centers and hormone secretion (Pang et al. 2013).

Experiments using goldfish and zebrafish eggs have revealed that the synthetic estrogen diethylstilbestrol (DES) directly induces oocyte maturation in fish (Tokumoto et al. 2004; Tokumoto et al. 2005). Therefore, we have focused on mPR as a molecule interacting with DES. A steroid binding experiment using the mPRα molecule expressed in cultured cells has proven that DES binds to the goldfish mPRα molecule (Tokumoto et al. 2007). This result is noted as the first clear demonstration of the nongenomic effects of an endocrine disruptor.

## Recent Advances Connecting cAMP and the Oocyte Maturation-Inducing Pathway

Immediately after the discovery of mPR, localization of mPR to the plasma membrane was investigated, and whether it was coupled to G proteins was determined. Localization to the egg surface was observed when mPR was expressed in oocytes as a GFP fusion protein (Hanna et al. 2006). Furthermore, immunoprecipitation experiments using mPR and G protein antibodies showed that mPR binds to Gi (Thomas et al. 2007). More recent studies have provided stronger evidence for the colocalization of mPR and Gi proteins on the plasma membrane using the proximity ligation analysis (PLA) method (Kelder et al. 2022). We found that DES is an agonist of oocyte maturation induction in goldfish and zebrafish and that DES also acts on mPR. We also showed that DES acts on mPR-responsive luminous cells transfected with goldfish mPR and modified luciferase and lowers the intracellular cAMP concentration in the same way as the maturation-inducing steroid DHP (Nakashima et al. 2015). In breast cancer cells expressing only human mPR, which does not show responsiveness to DES as an endogenous mPR, DES shows the same cAMP concentration-lowering effect as DHP. These results suggest that mPRα and Gi protein are coupled and are strong evidence of the coupling of Gi and mPR. Experiments with fluorescently labeled progesterone have also shown that DES inhibits progesterone binding to the cell surface (Figure 2).

These findings emphasize that the mechanism of fish oocyte maturation occurs through mPRα and Gi, as indicated in Figure 3. In fish in which MIH and mPR have been discovered, research on the coupling of mPR and Gi protein has progressed.

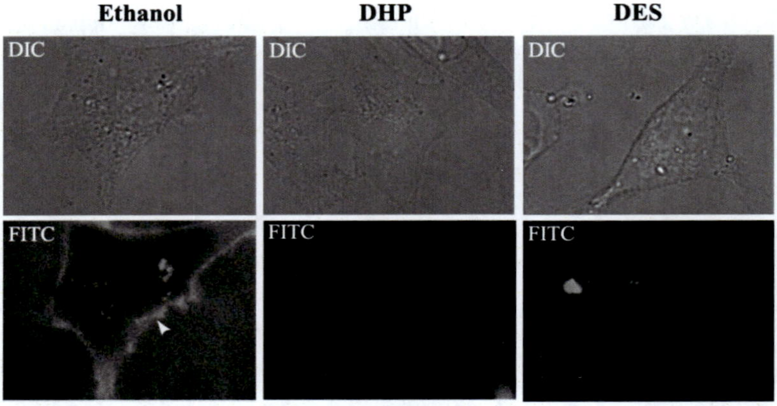


**Figure 2.** Inhibition of progesterone–BSA–fluorescein isothiocyanate (FITC) binding on the cell surface of breast cancer cells by 17,20β-DHP and diethylstilbestrol (DES). Differential contrast (DIC) images of goldfish mPRα gene transfected human breast cancer cells are indicated (upper panels). The fluorescent images were photographed in the presence or absence (ethanol) of compounds (lower panels). The cell surface FITC fluorescence observed in ethanol (white arrow head) disappeared with DHP and DES.

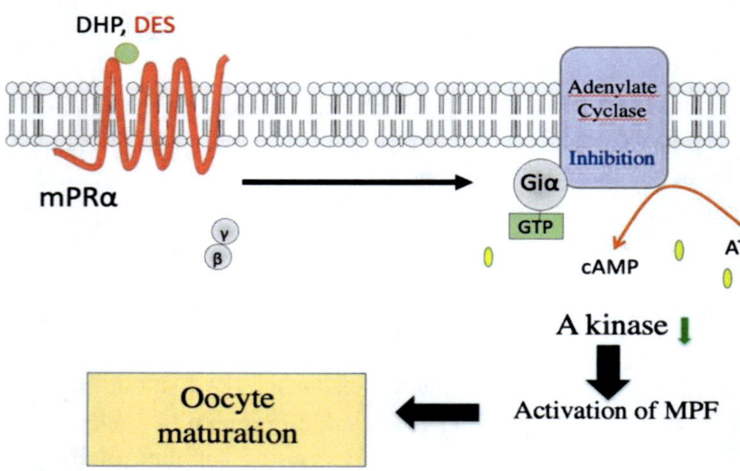


**Figure 3.** Induction mechanism of oocyte maturation in fish. Upon binding 17,20β-DHP or DES, Giα is activated and inhibits adenylate cyclase. This effect decreases the intracellular cAMP concentration, and A-kinase (cAMP-dependent protein kinase) activity is decreased, leading to activation of MPF and resulting in oocyte maturation.

However, the story has become diverse and controversial in frogs. As mentioned above, the decrease in intracellular cAMP concentration after treatment with progesterone led to the theory that a cell surface receptor coupled with a Gi pathway was involved. Since then, many experimental results have been published that both confirm and deny the Gi-mediated pathway derived from the detection of decreased intracellular cAMP concentration. Supporting the theory were findings that injection of the catalytic subunit of PKA (PKAc) inhibited progesterone-induced oocyte maturation (Daar et al. 1993; Eyers et al. 2005; Matten et al. 1994). Furthermore, treatment with phosphodiesterase (PDE) inhibitors (Andersen et al. 1998; Han et al. 2006) and CT suppressed maturation (Huchon et al. 1981; Mulner et al. 1979; Schorderet-Slatkine et al. 1978). Inhibition of AC promoted the maturation of progesterone-induced oocytes (Sadler et al. 1986). Consistently, injections of PDE, PKA regulatory subunit (PKAr), or PKA pseudo substrate inhibitor (PKI) were all found to reverse meiotic arrest (Andersen et al. 1998; Daar et al. 1993; Eyers et al. 2005). However, other studies failed to detect changes in cAMP levels in response to progesterone (O'Connor and Smith 1976; Schutter et al. 1975; Thibier et al. 1982). Inhibition of PKA activity using H-89, isoquinolinesulfonamide compound, or PKI injection failed to stimulate oocyte maturation (Noh and Han 1998). Induction of a decrease in cAMP level via stimulation of exogenously expressed Gi-coupled serotonin receptors did not stimulate oocyte maturation (Noh and Han 1998). Furthermore, a study using acetylcholine showed that reducing cAMP levels did not induce oocyte maturation (Gelerstein et al. 1988).

Based on these opposite reports, some different hypotheses for the induction of oocyte maturation in *Xenopus* have been proposed. One theory has emerged that in *Xenopus*, MIH reduces intracellular cAMP levels by inhibiting stimulatory G (Gs) proteins. The idea that a decrease in the concentration of intracellular cAMP is induced by inhibition of Gs proteins has been established (Gallo et al. 1995; Kalinowski et al. 2003). However, it is unclear how receptor–progesterone binding causes inhibition of Gs proteins. Surprisingly, it has been suggested that the ligand for MIH in frogs is not progesterone but testosterone (Hammes 2004). Additionally, experiments have shown that progesterone induces egg maturation not through the cell membrane receptor but through the proximity of the nuclear receptor to the cell membrane (Bayaa et al. 2000; Tian et al. 2000).

Research on the relationship between the receptor and G protein was terminated in the 1990s because of the length of time it took to discover the

MIH receptor. However, mPRβ has been shown to function as a progesterone receptor, although the possibility of coupling between mPRβ (paqr8) and G proteins has not been investigated (Josefsberg Ben-Yehoshua et al. 2007).

Recently, a new idea was suggested on the basis of research using intracellular FRET sensors: that the release of oocyte meiotic arrest in *Xenopus* is independent of a reduction in either cAMP level or PKA activity but rather proceeds through a parallel cAMP/PKA-independent pathway (Nader et al. 2016). The idea was based on a knockout study in mice in which knockout of G protein-coupled receptor 3 (Gpr3) led to premature oocyte maturation in antral follicles and sterility. The authors suggested the mechanism, in which the ortholog of gpr3 in *Xenopus* GPR185 maintains the high concentration of cAMP involved in meiotic arrest. However, the downstream mechanism regulated by mPRβ remains unclear. To clarify the mechanism, the binding of mPR and G proteins in frogs should be investigated.

## Conclusion

The experimental system of oocyte maturation has produced many results, providing a research target for the discovery of MPF and contributing to the cell cycle research that was initiated by that discovery. Ultimately, MPF research was awarded the Nobel Prize. The mechanism of steroid hormone induction, which is the initiation mechanism, was studied with several experimental methods, such as forced expression and inhibition via RNA injection, induction via drug treatment, and inhibition, and the general mechanism was established. However, research on the starting point of oocyte maturation ended prematurely because it took too long to find the receptor.

The discovery of mPRs was thought to be groundbreaking, but many researchers have already moved away from research on the reaction on the cell membrane, and research by the discoverer of mPR has been limited to the fish in which mPRs were discovered. However, quite surprising results have been published showing that zebrafish oocytes with knockout of 7 mPRs can still mature (Wu et al. 2020). Essentially, these results suggest that mPRs are not involved in the induction of oocyte maturation. However, whether there are any rescue pathways other than the MIH-mPR pathway in physiological states should be investigated. Additionally, the characteristics of the proteins expressed in the knockout strain should be analyzed. A system of frogs should also be tested to find a solid mechanism underlying these findings. It is believed that this will lead to an understanding of the nongenomic action of steroids centering on mPRs, which are expressed in cells throughout the body and are presumed to cause a wide range of physiological effects.

## Acknowledgments

This work was supported by Grants-in-Aid for Scientific Research on Priority Areas from the Ministry of Education, Culture, Sports, Science and Technology of Japan (JSPS KAKENHI grant number 20K06719 to TT). We also acknowledge a scholarship for foreign students from MEXT to Md. Forhad Hossain.

## References

Andersen CB, Roth RA and Conti M Protein kinase B/Akt induces resumption of meiosis in *Xenopus* oocytes. *J Biol Chem* (1998) 273, 18705-18708. doi: 10.1074/jbc.273.30.18705.

Bayaa M, Booth RA, Sheng YL and Liu XJ The classical progesterone receptor mediates *Xenopus* oocyte maturation through a nongenomic mechanism. *P Natl Acad Sci USA* (2000) 97, 12607-12612. doi: 10.1073/pnas.220302597.

Cicirelli MF and Smith LD Cyclic AMP levels during the maturation of *Xenopus* oocytes. *Dev Biol* (1985) 108, 254-258. doi:10.1016/0012-1606(85)90029-6.

Daar I, Yew N and Vande Woude GF Inhibition of mos-induced oocyte maturation by protein kinase A. *J Cell Biol* (1993) 120, 1197-1202. doi:10.1083/jcb.120.5.1197.

DeManno DA and Goetz FW The effects of forskolin, cAMP, and cyanoketone on steroid-induced meiotic maturation of yellow perch (*Perca flavescens*) oocytes in vitro. *Gen Comp Endocrinol* (1987a) 66, 233-243. doi:10.1016/0016-6480(87)90272-3.

Demanno DA and Goetz FW Steroid-induced final maturation in brook trout (Salvelinus fontinalis) oocytes in vitro: the effects of forskolin and phosphodiesterase inhibitors. *Biology of reproduction* (1987b) 36, 1321-1332.

Eyers, PA, Liu, J, Hayashi, NR, Lewellyn, AL, Gautier, J, & Maller, JL. Regulation of the G(2)/M transition in *Xenopus* oocytes by the cAMP-dependent protein kinase. *Journal of Biological Chemistry* (2005) 280, 24339-24346. doi: 10.1074/jbc.M412442200.

Gallo CJ, Hand AR, Jones TLZ and Jaffe LA Stimulation of *Xenopus* Oocyte Maturation by Inhibition of the G-Protein Alpha-S Subunit, a Component of the Plasma-Membrane and Yolk Platelet Membranes. *Journal of Cell Biology* (1995) 130, 275-284. doi:10.1083/jcb.130.2.275.

Gelerstein S, Shapira H, Dascal N, Yekuel R and Oron Y Is a Decrease in Cyclic-Amp a Necessary and Sufficient Signal for Maturation of Amphibian Oocytes. *Developmental Biology* (1988) 127, 25-32. doi:10.1016/0012-1606(88)90185-6.

Godeau JF, Schorderet-Slatkine S, Hubert P and Baulieu EE Induction of maturation in *Xenopus laevis* oocytes by a steroid linked to a polymer. *Proc Natl Acad Sci USA* (1978) 75, 2353-2357.

Haider S and Chaube SK Changes in total cAMP levels during oocyte maturation in the catfish, Clarias batrachus. *Comparative Biochemistry and Physiology Part A: Physiology* (1995) 112, 379-385.

Hammes SR Steroids and oocyte maturation—a new look at an old story. *Molecular Endocrinology* (2004) 18, 769-775.

Han, SJ, Vaccari, S, Nedachi, T, Andersen, CB, Kovacina, KS, Roth, RA, & Conti, M. Protein kinase B/Akt phosphorylation of PDE3A and its role in mammalian oocyte maturation. *EMBO J* (2006) 25, 5716-5725. doi:10.1038/sj.emboj.7601431.

Hanna R, Pang Y, Thomas P and Zhu Y Cell-surface expression, progestin binding, and rapid nongenomic signaling of zebrafish membrane progestin receptors α and β in transfected cells. *Journal of endocrinology* (2006) 190, 247-260.

Huchon D, Ozon R, Fischer EH and Demaille JG [Meiotic maturation of the ovocyte of *Xenopus laevis*: a 4-step mechanism]. *C R Seances Acad Sci III* (1981) 292, 65-67.

Ishikawa K, Hanaoka Y, Kondo Y and Imai K Primary action of steroid hormone at the surface of amphibian oocyte in the induction of germinal vesicle breakdown. *Mol Cell Endocrinol* (1977) 9, 91-100.

Iwamatsu T, Takahashi S, Sakai N, Nagahama Y and Onitake K Induction and inhibition of in vitro oocyte maturation and production of steroids in fish follicles by forskolin. *Journal of Experimental Zoology* (1987) 241, 101-111.

Jalabert B and Finet B Regulation of oocyte maturation in the rainbow trout, Salmo gairdneri: role of cyclic AMP in the mechanism of action of the maturation inducing steroid (MIS), 17α-hydroxy, 20β-dihydroprogesterone. *Fish Physiology and Biochemistry* (1986) 2, 65-74.

Josefsberg Ben-Yehoshua L, Lewellyn AL, Thomas P and Maller JL The role of *Xenopus* membrane progesterone receptor beta in mediating the effect of progesterone on oocyte maturation. *Mol Endocrinol* (2007) 21, 664-673.

Kalinowski RR, Jaffe LA, Foltz KR and Giusti AF A receptor linked to a Gi-family G-protein functions in initiating oocyte maturation in starfish but not frogs. *Dev Biol* (2003) 253, 139-149. doi:10.1006/dbio.2002.0860.

Kelder J, Pang Y, Dong J, Schaftenaar G and Thomas P Molecular modeling, mutational analysis and steroid specificity of the ligand binding pocket of mPRα (PAQR7): Shared ligand binding with AdipoR1 and its structural basis. *The Journal of Steroid Biochemistry and Molecular Biology* (2022) 219, 106082.

Maller JL, Butcher FR and Krebs EG Early effect of progesterone on levels of cyclic adenosine 3':5'-monophosphate in *Xenopus* oocytes. *J Biol Chem* (1979) 254, 579-582.

Masui Y and Clarke HJ Oocyte maturation. *Int Rev Cytol* (1979) 57, 185-282.

Masui Y and Markert CL Cytoplasmic control of nuclear behavior during meiotic maturation of frog oocytes. *J Exp Zool* (1971) 177, 129-145. doi:10.1002/jez.1401770202.

Matten W, Daar I and Vande Woude GF Protein kinase A acts at multiple points to inhibit *Xenopus* oocyte maturation. *Mol Cell Biol* (1994) 14, 4419-4426. doi:10.1128/mcb.14.7.4419-4426.1994.

Mulner O, Huchon D, Thibier C and Ozon R Cyclic AMP synthesis in *Xenopus laevis* oocytes: inhibition by progesterone. *Biochim Biophys Acta* (1979) 582, 179-184. doi:10.1016/0304-4165(79)90301-5.

Nader N, Courjaret R, Dib M, Kulkarni RP and Machaca K Release from *Xenopus* oocyte prophase I meiotic arrest is independent of a decrease in cAMP levels or PKA activity. *Development* (2016) 143, 1926-1936.

Nagahama Y and Adachi S Identification of maturation-inducing steroid in a teleost, the amago salmon (Oncorhynchus rhodurus). *Developmental biology* (1985) 109, 428-435.

Nakashima, M, Suzuki, M, Saida, M, Kamei, Y, Hossain, Md B, & Tokumoto, T. Cell-based assay of nongenomic actions of progestins revealed inhibitory G protein coupling to membrane progestin receptor α (mPRα). *Steroids* (2015) 100, 21-26.

Noh SJ and Han JK Inhibition of the adenylyl cyclase and activation of the phosphatidylinositol pathway in oocytes through expression of serotonin receptors does not induce oocyte maturation. *Journal of Experimental Zoology* (1998) 280, 45-56. doi: 10.1002/(Sici)1097-010x(19980101)280:1<45::Aid-Jez6>3.0.Co;2-H.

O'Connor CM and Smith LD Inhibition of oocyte maturation by theophylline: possible mechanism of action. *Dev Biol* (1976) 52, 318-322. doi:10.1016/0012-1606(76)90249-9.

Oba Y, Yoshikuni M, Tanaka M, Mita M and Nagahama Y Inhibitory Guanine-nucleotide-binding-regulatory Protein α Subunits in Medaka (*Oryzias latipes*) Oocytes: cDNA Cloning and Decreased Expression of Proteins During Oocyte Maturation. *European journal of biochemistry* (1997) 249, 846-853.

Pang Y, Dong J and Thomas P Characterization, neurosteroid binding and brain distribution of human membrane progesterone receptors delta and epsilon (mPRdelta and mPRepsilon) and mPRdelta involvement in neurosteroid inhibition of apoptosis. *Endocrinology* (2013) 154, 283-295. doi:10.1210/en.2012-1772.

Pellaz V and Schorderet-Slatkine S Evidence for a pertussis toxin-sensitive G protein involved in the control of meiotic reinitiation of *Xenopus laevis* oocytes. *Exp Cell Res* (1989) 183, 245-250. doi:10.1016/0014-4827(89)90434-5.

Sadler SE and Maller J Identification of a steroid receptor on the surface of *Xenopus* oocytes by photoaffinity labeling. *Journal of Biological Chemistry* (1982) 257, 355-361.

Sadler SE and Maller JL Inhibition of *Xenopus* oocyte adenylate cyclase by progesterone: a novel mechanism of action. *Adv Cyclic Nucleotide Protein Phosphorylation Res* (1985) 19, 179-194.

Sadler SE, Schechter AL, Tabin CJ and Maller JL Antibodies to the ras gene product inhibit adenylate cyclase and accelerate progesterone-induced cell division in *Xenopus laevis* oocytes. *Mol Cell Biol* (1986) 6, 719-722. doi:10.1128/mcb.6.2.719-722.1986.

Schorderet-Slatkine S and Baulieu EE Forskolin increases cAMP and inhibits progesterone induced meiosis reinitiation in *Xenopus laevis* oocytes. *Endocrinology* (1982) 111, 1385-1387. doi:10.1210/endo-111-4-1385.

Schorderet-Slatkine S, Schorderet M, Boquet P, Godeau F and Baulieu EE Progesterone-induced meiosis in *Xenopus laevis* oocytes: a role for cAMP at the "maturation-promoting factor" level. *Cell* (1978) 15, 1269-1275. doi:10.1016/0092-8674(78)90052-1.

Schutter AP, Kram R, Hubert E and Brachet J Cyclic nucleotides and amphibian development. *Exp Cell Res* (1975) 96, 7-14. doi:10.1016/s0014-4827(75)80030-9.

Stith BJ and Maller JL The effect of insulin on intracellular pH and ribosomal protein S6 phosphorylation in oocytes of *Xenopus laevis*. *Developmental biology* (1984) 102, 79-89.

Tang, YT, Hu, T, Arterburn, M, Boyle, B, Bright, JM, Emtage, PC, & Funk, WD. PAQR proteins: a novel membrane receptor family defined by an ancient 7-transmembrane pass motif. *J Mol Evol* (2005) 61, 372-380.

Thibier C, Mulner O and Ozon R In vitro effects of progesterone and estradiol-17 beta on choleragen activated *Xenopus* oocyte adenylate cyclase. *J Steroid Biochem* (1982) 17, 191-196. doi:10.1016/0022-4731(82)90121-2.

Thomas, Peter, Y Pang, J Dong, P Groenen, J Kelder, J de Vlieg, Y Zhu, C Tubbs. Steroid and G protein binding characteristics of the seatrout and human progestin membrane receptor α subtypes and their evolutionary origins. *Endocrinology* (2007) 148, 705-718.

Tian J, Kim S, Heilig E and Ruderman JV Identification of XPR-1, a progesterone receptor required for *Xenopus* oocyte activation. *Proc Natl Acad Sci USA* (2000) 97, 14358-14363. doi:10.1073/pnas.250492197.

Tokumoto M, Nagahama Y, Thomas P and Tokumoto T Cloning and identification of a membrane progestin receptor in goldfish ovaries and evidence it is an intermediary in oocyte meiotic maturation. *Gen Comp Endocrinol* (2006) 145, 101-108.

Tokumoto T, Tokumoto M, Horiguchi R, Ishikawa K and Nagahama Y Diethylstilbestrol induces fish oocyte maturation. *Proc Natl Acad Sci USA* (2004) 101, 3686-3690.

Tokumoto T, Tokumoto M and Nagahama Y Induction and inhibition of oocyte maturation by EDCs in zebrafish. *Reprod Biol Endocrinol* (2005) 3, 69.

Tokumoto, Toshinobu, Mika Tokumoto, Kumi Shimizuguchi, Tatsuya Fukuda, Etsuko Sugita, Manami Suzuki, Yu-ta Sakae, Yu-ichi Akiyama, Ryo Nakayama, Shimi Rani Roy, Md Saydur Rahman, Yefei Pang, Jing Dong, Peter Thomas. Characterization of multiple membrane progestin receptor (mPR) subtypes from the goldfish ovary and their roles in the induction of oocyte maturation. *General and comparative endocrinology* (2012) 177, 168-176.

Tokumoto T, Tokumoto M and Thomas P Interactions of diethylstilbestrol (DES) and DES analogs with membrane progestin receptor-alpha and the correlation with their nongenomic progestin activities. *Endocrinology* (2007) 148, 3459-3467.

Wu XJ, Liu DT, Chen S, Hong W and Zhu Y Impaired oocyte maturation and ovulation in membrane progestin receptor (mPR) knockouts in zebrafish. *Mol Cell Endocrinol* (2020) 511, 110856. doi:10.1016/j.mce.2020.110856.

Xu, J, Zeng, C, Chu, W, Pan, F, Rothfuss, JM, Zhang, F, Tu, Z, Zhou, D, Zeng, D, Vangveravong, S, Johnston, F, Spitzer, D, Chang, KC, Hotchkiss, RS, Hawkins, WG, Wheeler, KT, & Mach, RH. Identification of the PGRMC1 protein complex as the putative sigma-2 receptor binding site. *Nat Commun* (2011) 2. doi:ARTN 38010.1038/ncomms1386.

Yamauchi, T, Kamon, J, Ito, Y, Tsuchida, A, Yokomizo, T, Kita, S, Sugiyama, T, Miyagishi, M, Hara, K, Tsunoda, M, Murakami, K, Ohteki, T, Uchida, S, Takekawa, S, Waki, H, Tsuno, NH, Shibata, Y, Terauchi, Y, Froguel, P, … Kadowaki, T. Cloning of adiponectin receptors that mediate antidiabetic metabolic effects. *Nature* (2003) 423, 762-769. doi:10.1038/nature01705.

Yoshikuni M and Nagahama Y Involvement of an inhibitory G-protein in the signal transduction pathway of maturation-inducing hormone (17α, 20β-dihydroxy-4-pregnen-3-one) action in rainbow trout (*Oncorhynchus mykiss*) oocytes. *Developmental biology* (1994) 166, 615-622.

Zhu Y, Bond J and Thomas P Identification, classification, and partial characterization of genes in humans and other vertebrates homologous to a fish membrane progestin receptor. *Proceedings of the National Academy of Sciences* (2003) 100, 2237-2242.

# Biographical Sketches

## Md. Forhad Hossain

**Affiliation:** Integrated Bioscience Section, Graduate School of Science and Technology, Shizuoka University, Japan.

**Education:**

Md. Forhad Hossain received the B.Sc. and M.S. degrees in genetic engineering and biotechnology from the University of Rajshahi, Rajshahi, Bangladesh, in 2017 and 2018, respectively, and the second master's degree from the Department of Science, Shizuoka University, Shizuoka, Japan, in September 2021, where he is currently pursuing the Ph.D. degree with the Department of Bioscience, with a focus on ovulation-inducing genes of Zebrafish and recombinant expression, and characterization of the membrane receptor protein, under the Japanese Government [Ministry of Education, Culture, Sports, Science and Technology (MEXT)] Scholarship.

**Business Address:** Ohya 836, Suruga-ku, Shizuoka 422–8529, Japan

**Research and Professional Experience:**

From July 2017 to July 2019, he worked as a Research Assistant with the Molecular Biology and Protein Science Laboratory, University of Rajshahi, Rajshahi-6205, Bangladesh.

**Publications from the Last 3 Years:**

1. Toshinobu Tokumoto, Md. Forhad Hossain, Md. Maisum Sarwar Jyoti, Md. Hasan Ali, Md. Babul Hossain, Mrityunjoy Acharjee, Md. Rezanujjaman & Mika tokumoto. Two-Step Mechanism of cyclin B Degradation initiated by Proteolytic Cleavage with the 26 S proteasome in fish. *Scientific Reports,* June 2020. https://doi.org/10.1038/s41598-020-65009-w.
2. Rubel Rana, Forhad Hossain, Hasan Ali, Maisum Sarwar Jyoti, Toshinobu Tokumoto. Biochemical characterization of zebrafish Prss59.1. *Biochemical and Biophysical Research Communications,* May 2021. https://doi.org/10.1016/j.bbrc.2021.04.118.
3. Mrityunjoy Acharjee, Md. Hasan Ali, Md. Maisum Sarwar Jyoti, Md. Rezanujjaman, Md. Maksudul Hassan, Md. Rubel Rana, Md. Forhad Hossain, Shinya Kodani and Toshinobu Tokumoto. The antagonistic activity of Padina arborescens extracts on mPRα. *Natural Product Research,* August 2022. https://doi.org/10.1080/14786419.2022.2120873.
4. Md. Forhad Hossain, Md. Sohanur Rahman Sohan, Mahedi Hasan, Md. Masum Miah, Salek Ahmed Sajib, Sumon Karmakar, Khandaker Md. Khalid-Bin-Ferdaus, Ahmad Humayun Kabir, Md. Mamunur Rashid, Mamunur Rashid Talukder, Md Abu Reza. Enhancement of Seed Germination Rate and Growth of Maize (Zea mays L.) Through LPDBD Ar/Air Plasma. *Journal of Soil Science and Plant Nutrition,* January 2022. https://doi.org/10.1007/s42729-022-00771-6.
5. Md. Forhad Hossain, Mst. Ambia Akter, Md. Sohanur Rahman Sohan, Dr. Nigar Sultana, Md Abu Reza, Kazi Md. Faisal Hoque. Bioremediation potential of hydrocarbon degrading bacteria: isolation, characterization, and assessment. *Saudi Journal of Biological Sciences,* August 2021. https://doi.org/10.1016/j.sjbs.2021.08.069.
6. Md. Sohanur Rahman Sohan, Mahedi Hasan, Md. Jahangir Alam, Md. Forhad Hossain, Salek Ahmed Sajib, Md. Masum Miah, Md. Mahmudul Hasan Maruf, Khandaker Md. Khalid-Bin- Ferdaus, Ahmad Humayun Kabir, Md. Mamunur Rashid, Mamunur Rashid Talukder, and Md. Abu Reza. Enhancement of Seed Germination, Growth, Yield, and Nutritional Composition Using Low-Frequency Glow Discharge (LFGD) Plasma in Wheat (Triticum Aestivum). *IEEE Transactions on Plasma Science,* October 2022. https://doi.org/10.1109/TPS.2022.3188745.
7. Md. Sohanur Rahman Sohan, Mahedi Hasan, Md. Forhad Hossain, Salek Ahmed Sajib, Khandaker Md. Khalid-Bin-Ferdaus, Ahmad Humayun Kabir, Md. Mamunur Rashid, Mamunur Rashid Talukder, Mona M. Elseehy, Ahmed M. El-Shehawi and Md Abu Reza. Low-frequency glow discharge (LFGD) plasma treatment enhances maize (Zea mays L.) seed germination, agronomic traits,

enzymatic activities, and nutritional properties. *Chemical and Biological Technologies in Agriculture,* January 2022. https://doi.org/10.1186/s40538-021-00275-y.

8. Mahedi Hasan, Md. Sohanur Rahman Sohan, Salek Ahmed Sajib, Md. Forhad Hossain, Masum Miah, Md. Mahmudul Hasan Maruf, Khandaker Md. Khalid-Bin-Ferdaus, Ahmad Humayun Kabir, Mamunur Rashid Talukder, Md. Mamunur Rashid, Md. Moinuddin, Mona M. Elseehy, Ahmed M. El-Shehawi, Md Abu Reza. The Effect of Low-Pressure Dielectric Barrier Discharge (LPDBD) Plasma in Boosting Germination, Growth, and Nutritional Properties in Wheat. *Plasma Chemistry and Plasma Processing,* November 2021. https://doi.org/10.1007/s11090-021-10217-z.
9. Sumon Karmakar, Mutasim Billah, Mahedi Hasan, Sohanur Rahman Sohan, Md. Forhad Hossain, Kazi Md. Faisal Hoque, Ahmad Humayan Kabir, Md. Mamunur Rashid, Mamunur Rashid Talukder, Md Abu Reza. Impact of LFGD (Ar+O2) plasma on seed surface, germination, plant growth, productivity and nutritional composition of maize (Zea mays L.). Heliyon, March 2021. https://doi.org/10.1016/j.heliyon.2021.e06458.
10. Md. Sohanur Rahman Sohan, Mahedi Hasan, Md. Forhad Hossain, Salek Ahmed Sajib, Md. Masum Miah, Md. Asif Iqbal, Sumon Karmakar, Md. Jahangir Alam, Khandaker Md. Khalid-Bin-Ferdaus, Ahmad Humayan Kabir, Md. Mamunur Rashid, Mamunur Rashid Talukder, Md Abu Reza. Improvement of Seed Germination Rate, Agronomic Traits, Enzymatic Activity and Nutritional Composition of Bread Wheat (Triticum aestivum) Using Low-Frequency Glow Discharge Plasma. *Plasma Chemistry and Plasma Processing,* March 2021. https://doi.org/10.1007/s11090-021-10158-7.

## Toshinobu Tokumoto, PhD

**Affiliation:** Integrated Bioscience Section, Graduate School of Science and Technology, Shizuoka University

**Education:** PhD

Department of Molecular Biomechanics, The Graduate University for Advanced Studies. April, 1991-March, 1994 (conferred March, 1994).

Dissertation: "Oocyte Proteasomes: The structure and role in the regulation of meiotic maturation and fertilization in the goldfish, *Carassius auratus*." Supervisor: Dr. Y. Nagahama, Ph.D.

**Business Address:** Ohya 836, Suruga-ku, Shizuoka 422–8529, Japan

**Research and Professional Experience:**

Assistant Professor: Department of Biology and Geosciences, Faculty of Science, Shizuoka University

February, 1995-2002

Lecturer: Department of Biology and Geosciences, Faculty of Science, Shizuoka University

September, 2002-2004

Associate Professor: Department of Biological Science, Faculty of Science, Shizuoka University

October, 2004-2011

Professor: Department of Biological Science, Faculty of Science, Shizuoka University

February, 2011-present

Head of Integrated Bioscience Section, Graduate School of Science and Technology, National University Corporation Shizuoka University April, 2018-present

**Publications from the Last 3 Years:**

1. Atsuki Imai, Yoshihiko Hagiwara, Yuki Niimi, Toshinobu Tokumoto, Yumiko Saga, Atsushi Suzuki (2020) Dead end1 acts with Nanos2 and Nanos3 to regulate testicular teratoma incidence. *PLOS ONE* 15 (4) 1-18. April 27, 2020 https://doi.org/10.1371/journal.pone.0232047.
2. Toshinobu Tokumoto, Md. Forhad Hossain, Md. Maisum Sarwar Jyoti, Md. Hasan Ali, Md. Babul Hossain, Mrityunjoy Acharjee, Md. Rezanujjaman, Mika Tokumoto (2020) Two-Step Mechanism of Cyclin B Degradation Initiated by Proteolytic Cleavage with the 26S Proteasome in Fish. *Scientific Reports* 10:8924 June 2 IF 4.011 (2018) doi.org/10.1038/s41598-020-65009-w.
3. Md. Rezanujjaman, Razain Tanvir, Md. Hasan Ali and Toshinobu Tokumoto (2020) An agonist for membrane progestin receptor (mPR) induces oocyte maturation and ovulation in zebrafish in vivo. *Biochemical and Biophysical Research Communications,* 529 (2), 347-352. https://doi.org/10.1016/j.bbrc.2020.05.208.
4. Md. Mostafizur Rahaman, Ryo-ichi Kumagai, Toshinobu Tokumoto (2020) Rapid Induction of Female-to-Male Sex Change in Adult Zebrafish by Injection of an Aromatase Inhibitor. *Zebrafish* 17 (4), 261-267. August 13 http://doi.org/10.1089/zeb.2020.1864.
5. Theeranukul Pachoensuk, Taketo Fukuyo, Klangnurak Wanlada, Md. Rezanujjaman, Md. Mostafizur Rahaman, Kagura Sasaoka, Md. Maisum Sarwar Jyoti, Md. Rubel Rana, Md. Hasan Ali and Toshinobu Tokumoto (2020) Pax2a is expressed in oocytes and is responsible for early development and oogenesis in zebrafish. *Biochemical and Biophysical Research Communications,* 533 (3), 592-599. 10 December 2020 https://doi.org/10.1016/j.bbrc.2020.09.059.

6. Theeranukul Pachoensuk, Taketo Fukuyo, Md. Rezanujjaman, Klangnurak Wanlada, Chihiro Yamamoto, Akiteru Maeno, Md. Mostafizur Rahaman, Md. Hasan Ali and Toshinobu Tokumoto (2021) Zebrafish stm is involved in the development of otoliths and of the fertilization envelope. *Reproduction and Fertility* 2 (1) 7-16. Online Publication Date: Jan 2021. doi: https://doi.org/10.1530/RAF-20-0040 Online Publication Date: 16 Feb 2021.
7. Abdullah An Naser, Takehiro Miyazaki, Jun Wang, Shuji Takabayashi, Theeranukul Pachoensuk, Toshinobu Tokumoto (2021) MC4R mutant mice develop ovarian teratomas. *Scientific Reports* 11, 3483 1-10, 10.1038/s41598-021-83001-w.
8. Xinxin Wang, Xin Ma, Gaobo Wei, Weirui Ma, Zhen Zhang, Xuepeng Chen, Lei Gao, Zhenbo Liu, Yue Yuan, Lizhi Yi, Jun Wang, Toshinobu Tokumoto, Junjiu Huang, Dahua Chen, Jian Zhang, JiangLiu (2021) The Role of DNA Methylation Reprogramming during Sex Determination and Transition in Zebrafish. *Genomics Proteomics & Bioinformatics* February 2021 Feb 1 Accept doi: doi.org/10.1016/j.gpb.2020.10.004.
9. Md. Rubel Rana, Md. Forhad Hossain, Md. Hasan Ali, Md. Maisum Sarwar Jyoti and Toshinobu Tokumoto (2021) Biochemical characterization of zebrafish Prss59.1. *Biochemical and Biophysical Research Communications,* 560 (1), 32-36. 30 June 2021 https://doi.org/10.1016/j.bbrc.2021.04.118.
10. Md. Maisum Sarwar Jyoti, Md. Rubel Rana, Md. Hasan Ali and Toshinobu Tokumoto (2022) Establishment of a steroid binding assay for membrane progesterone receptor alpha (PAQR7) by using graphene quantum dots (GQDs). *Biochemical and Biophysical Research Communications,* 592 (1), 1-6. 12 February 2022 https://doi.org/10.1016/j.bbrc.2022.01.002.
11. Ali Md. Hasan, Md. Maisum Sarwar Jyoti, Md. Rubel Rana Md. Rezanujjaman and Toshinobu Tokumoto (2022) Purification and identification of the 20S proteasome complex from zebrafish. *Zebrafish,* Vol. 19, No. 1 18-23. 14 Feb 2022 https://doi.org/10.1089/zeb.2021.0064.
12. Mst. Habiba Mostari, Md. Mostafizur Rahaman, Mst. Afroza Akhter, Md. Hasan Ali, Tomohiro Sasanami, Toshinobu Tokumoto (2022) Transgenerational effects of bisphenol A on zebrafish reproductive tissues and sperm motility. *Reproductive Toxicology,* 109, 31-38. April 2022 Available online 2 March 2022 https://doi.org/10.1016/j.reprotox.2022.02.005.
13. Mrityunjoy Acharjee, Md. Hasan Ali, Md. Maisum Sarwar Jyoti, Md. Rezanujjaman, Md. Maksudul Hassan, Md. Rubel Rana, Md. Forhad Hossain, Shinya Kodani and Toshinobu Tokumoto (2022) The antagonistic activity of Padina arborescens extracts on mPRα. *Natural Product Research,* on line on 7 Sep 2022 https://doi.org/10.1080/14786419.2022.2120873.

# Chapter 7

# Looking Backwards and Forwards to the Rights of All Animals: Analogical Natural Law and 'Total Liberation'


## Michael Allen

East Tennessee State University, Johnson City, Tennessee, USA



## Abstract

Widely associated with humanity's dominion over the animals, natural law is a neglected resource in the discourse of animal rights. Animal rights theorists today show little interest in looking back to natural law to motivate liberating nonhuman animals from oppression by humans. However, analogical reasoning in natural law provides an alternative to univocal appeals to a singular criterion for all animals having rights to complex analogical continuities between differentiated, or 'unequal,' species and natures. Such reasoning also aligns in some considerable degree with the most forward-looking and radical approach to the liberation of all animals: total liberation. Despite reaching an impasse concerning the latter's anarchism, natural lawyers share with total liberationists an anthropocentric vision of highly variegated interspecies relations. Moreover, both insist upon the fundamental moral and political duty of thinking humans to liberate all animals from historical patterns and systems of oppression, unjustly constricting the fulfillment of diverse kinds, species, and natures. An alliance of natural law and total liberation promises to move the presently 'flat,' univocal discourse of animal rights forward in a fruitful and theoretically enriched direction.



**Keywords:** natural law, analogical reasoning, total liberation, cognitive ethology

In: Advances in Animal Science and Zoology. Volume 21
Editor: Owen P. Jenkins
ISBN: 979-8-89113-291-7

## Introduction

Natural law has a 'bad rap' in the present-day animal rights literature. It is usually associated by its critics with theological premises concerning God's grace and divine government of the cosmos. Looking backwards to medieval conceptions of natural law contributes little or nothing to the animal rights project, where this is viewed by some of its advocates as a desacralized or secular project of abolition and liberation. However, this speciesistic interpretation of natural law, depending ultimately on theological assumptions of grace and divine governance, is challenged by a more charitable reading of natural law emphasizing its analogical structure. On this alternative reading, medieval natural law entails analogical reasoning concerning the integration of all 'creatures' in complex anthropically ordered patterns and systems of continuity, similarity, or 'likeness,' between animal and human lives. Perhaps surprisingly, such an emphasis on analogical comparison and integration aligns with one of the most forward-looking and radical approaches to animal rights: 'total liberation.' This concerns the liberation of all animals – nonhuman and human – from human ordered systems and patterns of oppression. Such a liberation project of total liberation might well be thought to entail comparisons of human and animal lives to establish common interests of all animals in liberation.

Rather than natural law, total liberationist appeal to decidedly modern abolitionist movements for justice. These movements progress from the emancipation of human to 'animal slaves' from systemic patters of oppression. That said, theologically motivated advocates for analogical natural law would argue total liberationists – when theorizing an integrated scheme of nonhuman and human animal liberation – 'flatten out' any analogical hierarchy of species in medieval natural law. In this chapter, I deny that total liberation is 'flat;' univocal rather than analogical. On the contrary, I show that total liberation incorporates analogical reasoning about continuities, similarities, and likenesses between animals and human lives substituting cognitive ethology for God's grace and divine government. I make a cautious argument for a closer alignment between natural law and total liberation, showing how both frameworks may coherently and fruitfully inform one another.

I first look back to St. Thomas Aquinas's account of analogical reasoning across the multiple domains of law as response to natural law's bad rap in the present-day animal rights literature. I then look forwards to the prospect of a total liberation of all animals as envisioned by its leading theorist, Steve Best. Here I contrast Aquinas' natural law approach with Best's historical and ethological approach to the continuities of animal and human lives. Next, I consider a recent argument by the theologian, John Milbank, purporting to show that analogical natural law in Aquinas is incompatible with the secular historical and ethological orientations of liberationists like Best. Finally, I contend analogical natural law and total liberation can be united coherently and fruitfully.

## Looking Backwards to Natural Law

Natural law's 'bad rap' in the present animal rights literature is exemplified by Gary Chartier (2010). He reconstructs a footnote argument by John Finnis in his 1981 book *Natural Law and Natural Rights*. In this footnote, Finnis is not concerned with the natural moral status of animals, but rather arguing against treating so-called marginal case humans with limited capabilities as no different from nonhuman animals (henceforth, animals). Chartier reconstructs Finnis' argument concerning the *moral non-equivalency* of marginal case humans with animals thus:

1. Humans have rights.
2. If animals have rights, this must be in virtue of whatever it is in virtue of which humans have rights.
3. But humans and animals share nothing but sentience.
4. So, if animals were to have rights, this could only be in virtue of sentience.
5. But bare sentience cannot be a basis for right-claims.
6. So, animals do not have rights.

Chartier notes concerning 2) – in a footnote to his own article – that "Finnis does not articulate his assumption, but his argument seems to me to depend on it [for o]therwise, the claim that sentience is the only thing humans and animals with rights have in common would seem to be irrelevant" (2010, online). In this battle of the footnotes, something gets lost. This is any close attention to the analogical structure of natural law in St. Thomas Aquinas. I look back, then, to Aquinas' account of natural law. My aim is to show that the assumption 2) in Chartier's reconstruction of Finnis is misleading. To be sure, the conclusion to this reconstructed argument does not make sense without it. However, 2) can be interpreted through Aquinas's analogical account of natural law in a way that opens the possibility of a different conclusion: viz., animals *do* have rights.

Finnis's appeal to natural law does not fundamentally depend on concepts from the classic animal rights literature of the middle to late twentieth century (for example, Singer 1975, Regan 1984) such as sentience. That is, it does not depend on any single category of analysis as exclusively determining the application of natural law to kinds, essences, or species. Instead, Finnis appeals to four overlapping – or rather intersecting – "domains" of application in Aquinas's *Commentary on Aristotle's Ethics* (Finnis 2020, p. 39).

1. Patterns (and patterned things or subjects or inquiry) and systems that are what they are *quite independently of our thinking*.
2. Patterns or systems of thought by which we make our *thinking coherent* and fruitfully non-fallacious: logical.
3. The order (pattern) one can bring towards one's deliberating toward *acting to attain ends by means in the open horizon* of one's life (as individual or group)
4. The kind of pattern *one can impose upon* matter (stuff, materials, including one's bodily parts and operations) and the kind of *means-end practical rationality that concerns making (and working with) artefacts* in the widest usual sense, including arts, crafts and technologies or techniques including … language and its use.

Here Aquinas provides a vision of what the contemporary theologian John Milbank calls an "integrative politics" of all kinds, essences, and species "that would make no ultimate separation between the concerns of all natural creatures and of human beings …" (2021, pp. 1-2). In other words, such a politics of integration would effectively deny any ultimate separation between animals and human beings denying animals have (natural) rights. It would deny any such separation on grounds that the singular category of sentience is insufficient for rights (as in 5. of Chartier's reconstructed argument attributed to Finnis).

As Finnis himself puts it, each of the terms employed by Aquinas "'nature' ('of this nature'), 'essence' ('essentiality,' …), 'kind' ('of a different kinds'), 'identity' ('identical,' 'similar,' …), and so forth, is *correctly usable in each of the four domains*" (2020, p. 39). Indeed, Finnis goes on to say, "the precise meaning of each *shifts, more or less systematically, as its use shifts between domains*." Consequently, the various terms of natural law – nature, essence, and kind –are "*analogical terms*." They are "neither univocal nor merely equivocal" (pp. 39-40).

What does this mean for the possibility of animals having rights? Recognizing animals are what they are quite independently of our thinking (domain 1.), humans may quite 'coherently' think of them as having rights analogous to the rights of humans. Contrary to Chartier's reconstructed argument, this is a 'fruitfully non-fallacious,' indeed 'logical' conclusion (domain 2.). It brings something vital to our thinking about how to attain ends (the flourishing of all natures, essences, and kinds in our patterns or systems of thought) by means (of integrating them into these patterns or systems) in the 'open horizon' of our lives (domain 3.). Such a conclusion therefore 'imposes upon' animals (analogical, natural) rights as 'artefacts' of human 'language and its use' (domain 4.).

Clearly, then, natural law has been given a 'bad rap' in the present-day animal rights literature. Looking backwards to Aquinas' four domains' shows that such law – as itself an artefact (Finnis 2020) of analogical practical human reasoning about natures, kinds, and essences– can deliver an integrated vision of animal and human rights. To be sure, neither Aquinas nor Finnis explicitly argue for animals having rights. Nevertheless, and as affirmed explicitly by Milbank, this is a viable position for thinking humans to take based on the analogical, muti-domain, structure of natural law.

## Looking Forward to the Total Liberation of All Animals

Milbank's notion of an integrative politics is echoed in the radical program of 'total liberation' most forward looking approach to the rights of all animals. This is the radical program of 'total liberation' (Best 2014; also, Pellow 2014). Total liberationists today look forward to the liberation of all animals (and indeed all planetary life) from human systems of oppression through market driven natural resource exploitations, involuntary human and animal labor, and mass industrial food production (as detrimental to both humans and animals). However, they look forward by looking back not to Aquinas and natural but rather the eighteenth and nineteenth century radical movements to abolish human slavery. Here early campaigners for animals – like Henry Burgh and Ana Sewell – used the analogy of human and animal slaves consistently with natural law. That is, they shifted the terms of nature, essence, kind, and identity across domains to attain the ends of animal protection in the open horizon of political life of Boston, New York, and London (See my

Chapter 3 in Patterson et al., 2022; also see Freeburg 2020). However, appeal to a different theoretical framework total liberationist make no use of natural law and the method of analogy. They appeal to a different theoretical framework that captures some of the core features of natural law while challenging others.

As one of the key theorists of total liberation, Steve Best (2014) argues for a new abolitionism extending to all animal not only by appeal to Marx and Nietzsche but also radical feminist thinkers. What all have in common, according to him, is the view that power asymmetries – whether of capitalists over proletarians, master over slave moralities, men over women – are functions of how knowledge is socially constructed as power. This aligns with the natural law emphasis in Aquinas and Finnis on patterns and system of thought quite independent of patterns and system of nature. However, rather than focusing on analogies between the subjects of power asymmetries, Best prioritizes the 'standpoint' of the most oppressed kind in these asymmetries, in this case, the standpoint of oppressed animals. He develops this animal standpoint through an account of the co-evolutionary development of animals and humans. Such a co-evolutionary account might suggest a shifting of terms in the process of coevolution. Nevertheless, eschewing natural law method of analogy, Best uses animal-human coevolution to argue instead that enslaving and exploiting animals *precedes human slavery.* The gist of his coevolutionary story is that humans learned how to oppress and exploit other humans by first learning how to oppress and exploit humans. Further, humans learned how to justify oppressing and exploiting other humans by first learning how to justify oppressing and exploiting animals (Best 2014; also see my Chapter 7 in Patterson et al., 2022).

This grounds his appeal to the animal standpoint as the key to unlocking the pattern or system all oppressions in toto. Not despite but rather because of taking the animal standpoint as *analytically prior to* those of diverse oppressed human groups, total liberation offers an alternative mode of integrative politics. Like analogical natural law, it corrects an overemphasis on the singular relevance of sentience in favor of integrated view of the rights of all animals in patterns and systems of thought, and socially constructed knowledge, oppressing animals and subsequently humans themselves. Such an emphasis on patterns and systems of oppression points to the *liberation* of all natures and kinds, as opposed to their natural flourishing or well-being, across domains. This notion of liberation is further developed by other theorists who, contrary to Best himself, emphasize the agency of animals in facilitating liberation from oppression. Hence, animals may exercise political agency in acts of self-liberation or liberating other animals of different species or kinds (Allen and von Essen 2019). If correct, such claims of animal political agency in facilitating liberations generally – or in totality – radically challenge the implicit assumption of Chartier's reconstructed argument that bare sentience is insufficient for rights-claims. This is the implicit assumption in his reconstruction that such claims must be grounded in a kind of agency that is exclusive to humans, or, at least, non-marginal-case humans (Chartier 2010).

It is objected by Best and others (for example, Fotopoulos and Sargis 2006, Staudenmaier 2003) that the kind of political agency exhibited by animals is not the same as that exhibited by humans. Perhaps some animals are capable of isolated acts of self- or other liberations (Allen and von Essen 2019; also see Hribal 2013). However, only humans possess the higher order agential capacities for long-term strategic planning and organizing as necessary for a *movement* against the total system of all oppressions. Animals will therefore depend for their liberation from systemic oppression on humans as planners and organizers of an anthropocentric political movement. One response to this objection is to stress the oppression and liberation of animals nevertheless remains fundamentally "*linked* to human liberation, albeit through a relation of dependency, rather than equality" (Allen and von Essen 2019, p. 188). That is, they remain linked in the total pattern and system of human thought historically oppressing animals first but then oppressing humans. However, this points back again to natural law and its method of analogy, as I discuss next.

## A Theological Challenge to Desacralized Liberationism

Although not specially engaging with the literature on total liberation, the theologian John Milbank (2021) contends differences between the agential capacities of animals and humans can be interpreted through St. Thomas Aquinas' natural law method of analogy to ground an integrative politics of human-animal relations. Natural law emphasizes the "*analogical continuity* between humans and animals and [all] other creatures" (p. 2). Consistent with Aquinas's *Commentary on Aristotle's Ethics*, such continuity by analogy – across multiple domains of application – does not demand strict equality or 'identity' between the natures or essences of animals and humans but only 'similarity.' This similarity resides in their common fate as the "moral and political life" of humans is "naturally … shared with all other creatures" (8).

This might well be taken to suggest a relational natural law account (Moore 2001) of rights as broadly consistent with Donaldson and Kymlicka's (2011) account of the differential rights of animal citizens of a Zoopolis. Given a backdrop of the natural moral rights of all animals, Donaldson and Kymlicka argue citizenship rights may be jurisprudentially differentiated between different kinds and categories of animals depending upon how their lives are linked to those of human citizens in a pattern or system of social cooperation. Such a positive jurisprudential differentiation of rights would appear to capture the analogical continuity, and similarity of fate, concerning animal and human lives emphasized by Milbank.

Perhaps surprisingly, though, Milbank stops short of endorsing the unequal or differentiated rights of animals and humans, depending on their various species-natures and corresponding roles in an integral political scheme like Donaldson and Kymlicka's Zoopolis. His concerns here are theological. Indeed, Milbank defines his vision of integral politics as making "no ultimate separation between the concerns of all natural creatures and of human beings, nor between those of human animal nature *and human spiritual, grace endowed existence*" (pp. 1-2; my italics). It might be tempting to dismiss this theological appeal to 'grace' as unnecessary to apply coherently and fruitfully the multi-domain framework of natural law to animals. However, Milbank presents an historical argument of his own in defense of the necessity of grace for natural law. According to this historical argument, Aquinas' jurisprudential legacy was "wrecked" by Duns Scotus, among others. They saw natural law not as analogical but rather *univocal*. Such emphasis on univocality, or singularity of meaning, brought about a "disjoining of the different modes [or domains] of law" (p. 4). Once disjoined from Aquinas' multiple analogically linked domains, law ceases to be "something shared with other creatures" (p. 5).

Milbank further contends such a disjoining of modes created the condition of desacralized modernity, whereby "freedom itself is *abolished*" (p. 6). For, absent the analogical continuity of shared ends and fates between all creatures, there can be only a "pseudo-freedom" based upon an "endless assertion of new individual rights with no way of adjudicating competing claims" (7). However, this requires (non-marginal case) thinking humans to "seek unity" in God's grace; a unity that includes the "whole anthropically-ordered cosmos" (p. 7). Here Milbank quite explicitly equates the analogical structure of natural law as basis for adjudicating competing rights claims – in addition to any claims for liberation from historical patterns and systems of oppression – in Godless modernity with the "hierarchies of divine government" (p. 7).

Indeed, his position is the desacralization process of modernity not only univocalizes the language of rights and liberation but also 'flattens out' any analogical *hierarchy* of species and natures across law's domains: that is, different species and natures fundamentally *unequal* in their agential capacities for claiming rights and demanding liberation.

What might we make of this theological argument from Milbank for a return to natural law *as inseparable from* God's grace and divine government? Does this shut down any possibility of a coherent and fruitful alignment of analogical multi-domain natural law with total animal-human liberation? I contend that it does not. As I have already shown above, the key theorists of total liberation, like Best, do not actually flatten out natural hierarchies of agential capacities between animals and humans into univocal rights claims to liberation. On the contrary, they acknowledge an anthropocentric order of animal-human relations, as relations of dependency rather than 'flat' equality, in the total liberation struggle. This "critical anthropomorphism" is a not function of divine government, but instead "cognitive ethology:" the close scientific observation of animal behaviors, and the "emotional, cognitive, and social complexity" of animal lives (Best 2014, p. 121; also see Griffin 2001). Does such a scientific appeal to cognitive ethology capture the multi-domain structure of natural, along with its analogical hierarchies of species and natures, *without* Milbank's appeal to divine government? A case can be made that it does.

Cognitive ethology analogizes the cognitive, emotional, behavioral, and social worlds of animals and humans. Thinking humans may say, coherently and fruitfully, that "Michael the gorilla loved the music of Luciano Pavarotti, just as Flint the chimp died of grief upon the death of his mother, Flo" (Best 2014, p. 123). Ethological observations allow such humans to say, coherently and fruitfully, that many animals "mourn their dead, enact burial rituals … know joy as well as sorrow … possess an aesthetic sense, and a sense of humor." Furthermore, many animals themselves have abilities in "analogical reasoning" in addition to "counting, memory, spatial mapping … [and] … problem solving." Many "evince a clear sense of morality, justice, and fair play" (pp. 124-5). All these ethologically grounded observational claims may appeal *analogically* to the shared ends and fates of all animals in the anthropically ordered cosmos. Scientific and critically anthropocentric, this does not 'flatten' animal human relations but rather acknowledges an analogical hierarchy of ends and fates across species and domains. Such an analogical understanding of such similarities between different species, kinds, and natures need not be sought through God's grace, any more than this analogical order must be attributed to divine government.

If at all persuasive in showing natural law *can be separated* from grace and divinity, contrary to Milbank, then these remarks perhaps open the prospect of looking forward to an alignment of natural law with total liberation. Rather than 'abolishing freedom,' such an alignment of frameworks abolishes patterns and systems of oppression. It abolishes patterns and systems based on humans narrow-mindedly – or univocally – failing to engage complex analogical hierarchies and continuities between different animal and human lives: emotionally, cognitively, morally, and so on.

## Backwards, Forwards Together?

Might, then, the two frameworks of natural law and total liberation come together coherently and fruitfully to inform one another about the prospects for liberating animals and human alike from anthropic patterns and systems of oppression? I answer cautiously in the affirmative. Contrary to its flat, univocal variant instigated by Duns Scotus, Aquinas' analogical natural law does not 'abolish freedom.' On the contrary, it embraces a conception of freedom as fulfilling multiple natural life-purposes – cognitive, emotional, social, moral – across the different domains of law. This requires that anthropically ordered patterns and systems of animal-human relations facilitate such a complex vision of multi-variant, multi-species freedom. The anthropic – or, in Best's language, critically anthropocentric – order should ensure patterns and systems do not oppress the species or natures of any animals. In other words, it should aim to 'liberate' them all.

What does analogical natural law get from this? It gets a considerable boost in its credibility within animal rights, and animal liberation, circles. Chartier's reconstructed argument is univocally 'flat' from the perspective of analogical natural rights. It considers only sentience as basis for animals having rights. However, consideration of the domains of natural law provides a basis for saying all animals have natural rights against systemic oppression based on multiple analogical factors extending well beyond mere sentience. Not only does it provide a credibility boost to natural law but also it helps to push the animal rights-liberation movement beyond its present-day malaise. Entirely flat questions about whether animals have rights univocally 'in virtue of whatever it is in virtue of which humans have rights' result only in theory fatigue. They result in exhaustion and frustration with theory, along with a corresponding loss of impetus and direct for the movement, precisely because *there is no* plausible singular, univocal answer (see my Chapter 7 in Patterson et al., 2022). Emphasizing the complexity of animal lives, total liberationist, like Best, recognize the theoretical poverty of univocal reasoning about rights. However, liberationists might also get a boost in credibility for their program by acknowledging the analogical method of natural law, in Aquinas's *Commentary*, as an overlooked legacy resource contributing theoretically to reenergizing an otherwise stagnant movement for liberation.

Indeed, those modern abolitionist movements against human and animal slavery that Best claims as legacy for his new abolitionism were often religiously inspired (see my Chapter 3 in Patterson et al., 2002). However, in the case of northern American abolitionism, these motivations had less to do with natural law and analogical reasoning than pathos, as empathetic basis for both religious experience and the rights of all suffering others (Ibid; also see Clerk 1995). Such eighteenth and nineteenth century religious sentimentalism did not endure into the late twentieth century when academic philosophers – rejecting sentimentalism – reformulated animal rights based on rigorous argument from univocal criteria like sentience (for example, Singer 1975). Acknowledging and embracing a natural law legacy in the struggle to liberate all animals, liberationists may offer believers a reentry point into the radical abolitionist tradition from which they had been alienated by its subsequently adopting a narrow secular approach to rights. To be sure, as theorized by Best, the new abolitionism is a desacralized movement, substituting science for God. Provided they are willing to set aside ideological differences concerning the divine origin and government of the cosmos, both religious and secular camps may 'seek [integral political] unity' in a renewed, reinvigorated movement for abolition and liberation.

That said, even if it is granted such an alliance of analogical natural law and total liberation points a way forward from flat theorizing resulting only in fatigue and enervation, difficult questions arise for such an alliance at the level of movement strategy. Natural lawyers have always maintained a close relationship between natural and human law. Natural law should inform and guide positive legislation. This might appear to place natural law within the legal reformist branch of the animal rights movement (for example, Wise 2000), exemplified, for example, by the Humane Society of the United States. Reformism stays within the boundaries of law. emphasizing lobbying elected representatives or organizing consumer boycotts against products based on the exploitation and abuse of animals by corporations. However, total liberationists distinguish themselves from this legal reformist branch, endorsing alternative extra-legal strategies for liberation ranging from nonviolent civil disobedience to overt intimidation tactics, like confronting animal abusers at their homes, sending them threatening letters, graffitiing their workplaces, or posting their names and addresses on websites (Allen and von Essen 2018).

Does this mean that any alliance of analogical natural law with total liberation breaks down over legal reformism as opposed to disobedience, or even more radical strategies and tactics of intimidation? Not necessarily. Aquinas permits civil disobedience in the case of tyrannical, unjust human laws. Contravening natural law, unjust human laws are not properly law at all, and so they cannot command obedience. On the contrary, disobedience in the face of unjust human law is not just an option, but rather a duty because such laws commit a serious offense against God (*Summa*, p. 1998). This allows for strategic alliances between natural lawyers and total liberationists concerning nonviolent disobedient challenges to injustices of oppression and exploitation. However, does Aquinas's duty to disobey unjust human laws also underwrite the more radical strategies of confrontation and intimidation? That might seem a lot less likely. Confrontation and intimidation are not strategies of disobedience to unjust human laws, but rather "ethical vigilantism." This consists of the "virtuous" taking the law into their own hands by "forcing the choices" of unjust individuals and "enforcing new norms" (Allen and von Essen 2018, p. 43).

Such vigilantism might well be thought to imply a natural law framework insofar as its prescriptions of new norms liberating all animals from oppression originate in an anarchistic vision of human thought detached from human law. These must therefore be *natural legislative judgements* concerning the analogically shared cognitive, emotional, social, and moral ends and purposes of all animals. This is consistent with Acquinas' emphasis on checking the moral deviancy of individual natural law breakers. Nevertheless, the anarchist orientation of total liberationists seeking to change behaviors of abusive individuals without appeal to the mechanism of human law also detaches from Aquinas' insistence upon *deriving human from natural law*, as means to achieving the shared or common good. In addition to concerns over the liberation tactics themselves crossing a line into abusiveness, the fundamental Aquinian objection to ethical vigilantism is that it narrows the scope of natural law to restructuring norms. In other words, it unacceptably narrows the scope of analogical reasoning about the continuities between animal and human lives to restructuring norms when the proper goal of natural law to achieve the good for all creatures by restructuring human law.

## Conclusion

Despite reaching an impasse with total liberationist concerning their more radical anarchistic orientation to bypassing legislative reforms, natural lawyers may align themselves with the former's project of liberating all animals from oppression and exploitation through patterns and systems of thought and action whereby humans order their lives in relation to animals. Why does such an alignment of frameworks matter? Combining religious and secular perspectives, such an alliance promises to move the presently 'flat' discourse of animal rights based on univocal appeals to a singular criterion for all animals having rights to complex analogical continuities between otherwise differentiated, or 'unequal,' species and natures. Indeed, both natural lawyers and total liberationist embrace the inescapable anthropocentrism of highly variegated interspecies relations, and the necessity for thinking humans to mobilize the abolitionist movement for liberation. This is compatible with acknowledging some animals have cognitive, and even moral, capacities well beyond mere sentience. These capacities are sufficient for them to engage in acts of self and other liberation. However, advocates for both frameworks also acknowledge that mobilizing a social movement to liberate all animals from historical patterns and systems of oppression unjustly constricting the fulfillment of diverse species and natures necessarily remains a fundamentally human duty.

If animal rights may be said to have a past in the analogical reasoning of medieval natural law, then looking backward to reclaim this legacy, and aligning it with the secular project total liberation, points to the possibility of a new way forward for animal rights.

## Disclaimer

None.

## References

Allen M., von Essen E. Interspecies Political Agency in the Total Liberation Movement. *Between the Species* (2019) 23(1), 175-199.

Allen M. Legacy: Overcoming Confusion and Fatigue. *Rethinking the American Animal Rights Movement*, Patterson-Kane E., Allen M., Edie, J. (Routledge, 2022).

Allen M. Origins of the Movement: Empathy and Emancipation, *Rethinking the American Animal Rights Movement*, Patterson-Kane E., Allen M., Edie, J. (Routledge, 2022).

Aquinas T. *New English Translation of St. Thomas Aquinas's Summa Theologiae*, tr. Fredosso A. (University of Notre Dame, www.nd.edu/-afreddos/summa-translation/TOC.htm).

Best S. *The Politics of Total Liberation* (Palgrave Macmillan, 2014).

Chartier G. Natural Law and Animal Rights. *Canadian Journal of Law and Jurisprudence* (2010) 33.

Clerk E. The Sacred Rights of the Weak: Sympathy and the Culture of Individual Rights in Antebellum America. *Journal of American History* (1995) 83: 467-493.

Cudd A. How to Explain Oppression: Criteria of Adequacy for Normative Explanatory Theories. *Philosophy of the Social Sciences* (2005) 35(1): 20-49.

Donaldson S., Kymlicka W. *Zoopolis*. (Oxford: Oxford University Press, 2011).

Finnis J. *Natural Law and Natural Rights*.

Finnis J. The Nature of Law. *Cambridge Companion to the Philosophy of Law*, ed. Tasioulas J. (Cambridge University Press, 2020).

Fotopoulos T., Sargis J. Human Liberation vs. Animal Liberation, *International Journal of Inclusive Democracy* (2006) 2(3).

Freeburg E. (2020). *A Traitor to His Species: Henry Burgh and the Birth of the Animal Rights Movement*. (New York: Basic Books, 2020).

Griffin D. *Animal Minds: Beyond Cognition to Consciousness*. (Chicago: University of Chicago Press, 2001).

Hribal J. *Fear of the Animal Planet: The Hidden History of Animal Resistance*. (Petrolia CA: Counterpunch, 2011).

Milbank J. Aquinas' Natural Law Isn't Only About Human Beings. *Church Life Journal* (2021).

Pellow D. *Total Liberation* (University of Minnesota Press, 2014).

Singer P. *Animal Liberation*. (Harper, 1977).

## Chapter 8

# The Effect of Urbanization on Carabidae and Scarabaeidae Families in Eskişehir Osmangazi University's Meşelik Campus and Meşelik Forest Gradient

**Ebru Ceren Fidan***, **PhD**
**D. Ümit Şirin, PhD**
**and Hilal Deniz Çelik, BD**
Department of Biology, Eskişehir Osmangazi University, Eskişehir, Türkey

## Abstract

With this study, the effect of urbanization on the Carabidae and Scarabaeidae families was studied in Eskişehir Osmangazi University Meşelik Campus and Meşelik Forest gradient. 3 different areas were chosen for the sampling. These areas are: Natural Habitat, Semi Natural Habitat and Urbanization area. Specimens were collected between May 2017 and May 2018 with 15-day periods and pitfall trap collection methods. Habitat similarities are determined by Sørensen and Jaccard indexes. Habitat diversities are determined by Shannon-Wiener and Simpson's indexes. As a result of the study, 12 different species from the Carabidae family and 7 different species from the Scarabaeidae family were determined. According to the results of the similarity index, Semi Natural Habitat and Natural Habitat were determined as the most similar areas with Jaccard's index with 0,368 value and Sorensen's index with 0,538 value. According to the diversity index, the Urbanization region is the poorest area in terms of species diversity, but this region where the

* Corresponding Author's Email: ebruceren@ogu.edu.tr.

In: Advances in Animal Science and Zoology. Volume 21
Editor: Owen P. Jenkins
ISBN: 979-8-89113-291-7

highest values are observed in terms of the number of individuals of the existing species.

**Keywords**: Carabidae, Scarabaeidae, urbanization effect, diversity, index

## Introduction

Today, with the expansion of urban areas worldwide, there has also been a significant increase in population. In the last century, 2% of the world's population lived in urban settlements, whereas today this rate has exceeded 15% of the population. In developed regions, including Western Europe, the USA and Japan, the proportion of people living in urban centres is approximately 45% compared to rural areas. As such, rapid urbanization and its consequences have had a global impact. In recent years, the concept of urbanization has been examined, including many aspects which affect people's lives such as crime, health, pollution, socio-economic, trade, education and market conditions. All of them are nowadays defined as important problems parallel to growth and crowding (Tanrıkulu, 2006).

In addition to the increase in population, human activities have also led to significant changes in the utilisation of the natural environment. This includes intensive building construction that covers city surfaces, involving the use of materials such as concrete, asphalt, brick and stone. Additionally, there has been an increase in industrial activities, vehicles and traffic which has resulted in higher fuel consumption and pollution.

As the urban area grows, the urban climate also differentiates compared to the surrounding rural climate. It is known that every building in a city, every factory, power plant and every source that releases chemicals into the air affects the atmosphere. This effect can greatly disturb the balance between the earth and the air and can have a negative impact. For example, according to the United Nations Environment Programme, the use of Methyl Bromide affects ozone depletion by 5% to 10% (Spinney, 1995).

In developed countries, the factor of urbanization affecting the variety and richness of species and systems in the natural environment has been revealed and discussed in many studies (Samways, 1989; Morse, 2001; Doygun and Alphan, 2006; Acar et al., 2007; Atmış et al., 2007; McKinney, 2008; O'Driscoll, 2010). The effect of urbanization has been studied on insects as well as many other groups. It is observed that the majority of studies focuses on the Coleoptera group.

The individuals of the Carabidae, Scarabaeidae, Leididae, Curculionidae and Staphylinidae families of Coleoptera are the most frequently used taxa for monitoring environmental changes. Sphenophorus of Curculionidae and Gonocephalum rusticum of Tenebrionidae are very sensitive to grazing and human activities. They decrease in numbers of individuals in environments where these activities are carried out. (Cano and Schuster, 2006; Nahmani et al., 2006). Carabids and Scarabaeids are well known in taxonomic and ecological terms. Carabids are distributed over a very wide geographical area and large habitats, except in the driest parts of deserts. Carabids consist of geographically widespread, ubiquitous species (ubiquist species), occupying large habitats (eurytopic), and species specialised to one or a few habitats (stenotopic) (Pearsall and Dunsworth, 2003; Rainio and Niemela, 2003). Both groups have species and genera that are specialised to specific habitats and living conditions. Therefore, they are used in ecological studies as bioindicators (Pearce and Venier, 2006; Shahabuddin et al., 2014; Syaripuddin et al., 2015).

Numerous examples of studies investigating the effects on insect groups and habitats can be found (Alaruikka et al. 2002; Rickman and Connor, 2003; Venn et al., 2003; Zanette et al., 2005; Magura et al., 2008; Girgin et al., 2010).

As it can be observed in the studies, habitat loss or habitat fragmentation caused by urbanization generally leads to the downsizing of populations living in natural areas, the fragmentation of populations and thus the distribution, and the decrease in species variety and diversity in habitats. In the present study, a survey sample was taken from Eskisehir Osmangazi University Meşelik Campus and Meşelik Forest, and it was aimed at measuring the reaction of human impact on Carabidae and Scarabaeidae gradually in three main areas.

As in previous studies, the study areas were selected as a gradient of

- Natural habitat (oak forest);
- Semi-natural habitat (campus and forest border zone) and
- Urbanization zone (campus zone).

By this way;

- The species composition of the collected species was determined,
- Information was gained about their ecology and habitat preferences,
- Comments could be made on the impact of urbanization and human impact on species.

# Material and Method

## Research Areas

The study area is Eskisehir Osmangazi University Meşelik Campus and Meşelik forest. The area is located between 39° 44° North parallels and 30° 29° East meridians. It has a surface of 50 hectares with a density of pine and an altitude between 850-1050 metres above sea level (Fidan et al., 2014). Three study areas were identified, which were thought to best represent the different habitats present in the Meşelik forest (Figure-1). These areas are;

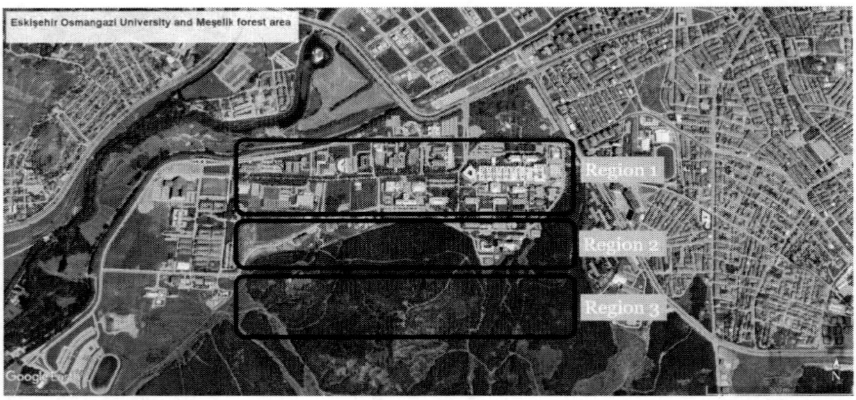


**Figure 1**. Image of the selected areas on the satellite photo of the Meşelik Campus and Meşelik forest.

*Region 1:* Urbanization zone: Campus Zone; The active zone of Eskisehir Osmangazi University Meşelik Campus in terms of building, road and human density (Figure 2).

*Region 2:* Semi-Natural habitat: Campus and forest border zone; The border zone between the Campus and Oak forest where both urbanization and natural area are intertwined (Figure 3).

*Region 3:* Natural habitat: Oak forest; It is a natural habitat away from human influence. Each area was sampled by approximately 500 m2 in diameter (Figure 4).

**Figure 2.** Region 1; Urbanization zone.

**Figure 3.** Region 2: Semi-Natural habitat.

**Figure 4.** Region 3: Natural habitat.

## Sampling Method

Samples were collected in 15 day periods for 12 months between May 2017 and May 2018. The beetle sampling was carried out with pitfall traps and by hand. 15 pitfall traps were placed (20 m intervals) to each sampling region. The pitfall traps (10 cm diameter x 15 cm depth, hard plastic cups) were burried in the soil at ground level and the trap was left in the ground for two weeks. A mixture of vinegar and vine was used as preservation liquid in traps. Many of the collected specimens stored in %70 alcohol while some of them were pinned for identification. They become museum material using the traditional preparation tecniques. All specimens are kept in the Entomology Laboratory of Eskişehir Osmangazi University. The specimens were separated and identified under a Leica MZ16 stereo microscope using identification keys (Hurka, 1996; Avgın, 2006; Löbl and Smathana, 2004; Şenyüz, 2009).

## Diversity and Similarity Analysis Methods

The Multi Variate Statistical Package (MVSP) (Kovach, 2009) was used to compare the diversity and similarity parameters of the insects. The frequency/abundance of the Carabidae and Scarabaeidae species from regions and the diversity and similarty of the areas were compared.

Frequency assessment was calculated by the formula given below (by Krebs, 1994);

Total number of samples in which a species occurs

Frequency percentage = x100 ———————————————

Total number of samples conducted during the study

The abundance of the species was calculated and grouped with Trotuş and Naie (2008).

- The rarest species < 1.1%
- Rare species = 1.2-2.0%
- Less dominant species = 2.1-5.0%
- Dominant species = 5.1-10.0%
- Most dominant species > 10.1%

Sørensen and Jaccard similarity indices were used to determine the faunal similarity relationships between the regions (Krebs, 1999);

Sørensen index;

$$S = \frac{2a}{2a + b + c}$$

where; a: the number of species in the first commune,
b: number of species in the second commune,
c: the number of common species in both communes.

Jaccard index;

$$C_j = \frac{a}{a + b + c}$$

where; $C_j$ = Jaccard's similarty index
a = the number of common species in both areas
b = the number of species inthe first are but not in the second
c = the number of species in the second are but not the first.

$C_j$, takes a value between 0 and 1. A value of (0) indicates that there are no common species in both communuties, while a value of (1) indicates that all species present in both communities (Jarvis, 2000; Magurran, 2004).

Shannon-Wiener and Simpson's indexes were used to determine species diversity (Krebs, 1999):

*Shannon-Wiener Diversity Index;*

$$H' = \sum_{i=1}^{s} (p_i)(\log_2 p_i)$$

Here;

$p_i$; the ratio of the number of individuals of the i-th species to the number of individuals.

ln; denotes the base of natural logarithm.

The $H^1$ value usually varies between 1.5 and 3.5. Rarely exceeds 4.

*Simpson's Diversity Index;*

$$D = \sum p_i^2$$

Simpson's index is based on the probability that two individuals randomly selected from a community belong to the same species. Therefore, as the value of D increases, diversity will decrease. Fort his reason, the index is usually used as 1-D or 1/D. Thus, an increasing value of D will indicate higher diversity.

## Results

Anatolia is one of the richest countries in Europe and the Middle East in terms of biological diversity. European fauna approaches 60,000 species, while this number reaches 75,000 in our country (Şekercioğlu et al., 2011). Climate differences, geological and geomorphological diversity, different environments such as sea, rivers, lakes and elevation differences in our country are the reasons for our biological richness. Due to that, many species are able to find a unique habitat. The province of Eskişehir, which is the study area, is preferred by many species because of its ecological conditions; this region has both microclimatic Mediterranean and continental climates.

Due to their variety in habitat preferences and their importance in the food chain, many faunistic and ecological studies have been carried out on ground beetles (Carabidae) and May beetles (Scarabaeidae) in Turkey (Fidan and Şirin, 2016). However, so far, the effects of urbanization on the species diversity of these groups in Eskisehir have not yet been investigated in detail. In our country, the species composition of these groups and their role in environmental assessments needs to be discussed. This study is the first of its kind to determine the fauna of Carabidae and Scarabaeidae in Eskişehir Osmangazi University Meşelik Forest and to evaluate the effects of urbanization on these groups.

A total of 390 individuals were collected from the areas selected for the study. A total of 362 individuals belonging to 12 different species of the Carabidae family and 28 individuals belonging to 7 different species of Scarabaeidae family were identified. The identified species, the areas where they were detected, the number of individuals and the frequency of the species in the area are shown in Table-1. All of the identified species are known from Eskişehir province.

**Table 1.** Species, Specimen number, and Species frequency

| | Region 1 Urbanization Zone | Region 2 Semi-Natural Habitat | Region 3 Natural Habitat | Total | Frequency |
|---|---|---|---|---|---|
| CARABIDAE | | | | | |
| *Calathus (Neocalathus) cinctus Motschulsky, 1850* | 1 | 9 | 68 | 78 | 20 |
| *Calathus (Calathus) fuscipes (Goeze, 1777)* | 1 | 79 | 129 | 209 | 53 |
| *Cicindela (Cicindela) campestris Linne, 1758* | 2 | 2 | 7 | 11 | 2,8 |
| *Carabus (Pachystus) graecus Dejean, 1826* | 0 | 13 | 17 | 30 | 7,7 |
| *Calathus (Neocalathus) melanocephalus (Linne, 1758)* | 0 | 11 | 4 | 15 | 3,8 |
| *Calosoma (Calosoma) sycophanta (Linne, 1758)* | 0 | 0 | 1 | 1 | 0,25 |

**Table 1.** (Continued)

| | Region 1 Urbanization Zone | Region 2 Semi-Natural Habitat | Region 3 Natural Habitat | Total | Frequency |
|---|---|---|---|---|---|
| *Harpalus (Harpalus) distinguendus (Duftschmid, 1812)* | 0 | 0 | 1 | 1 | 0,25 |
| *Ophonus (Hesperophonus) azureus (Fabricius, 1775)* | 0 | 0 | 2 | 2 | 0,5 |
| *Amara (Amara) eurynota* (Panzer, 1796) | 0 | 0 | 1 | 1 | 0,25 |
| *Calathus (Neocalathus) erratus (C.R. Sahlberg, 1827)* | 0 | 0 | 2 | 2 | 0,5 |
| *Pterostichus (Platysma) niger (Schaller, 1783)* | 0 | 8 | 2 | 10 | 2,5 |
| *Cymindis (Cymindis) lineata (Quensel in Schonherr, 1806)* | 0 | 2 | 0 | 2 | 0,5 |
| SCARABAEIDAE | | | | | |
| *Cetonia aurata subsp. pallida* (Drury, 1770) | 0 | 0 | 1 | 1 | 0,25 |
| *Oxythyrea cinctella ab. natalia* (Olsoufieff, 1916) | 0 | 0 | 3 | 3 | 0,8 |
| *Protaetia (Netocia) vidua* (Gory & Percheron, 1833) | 0 | 0 | 1 | 1 | 0,25 |
| *Protaetia (Netocia) cuprea* (Fabricius, 1775) | 0 | 4 | 8 | 12 | 3 |
| *Holochelus (Miltotrogus) sp.* Reitter, 1902 | 0 | 0 | 7 | 7 | 1,8 |
| *Oxythyrea cinctella (Schaum, 1841)* | 1 | 0 | 1 | 2 | 0,5 |
| *Netocia (s.str.) hungarica anatolica Medvedev, 1947* | 0 | 0 | 2 | 2 | 0,5 |

Calathus (Calathus) fuscipes (Goeze, 1777) has the highest number of individuals among all detected species with 209 individuals. This species is seen in all sampling areas even if it has a low number of individuals. In both families, it can be concluded that the specimens prefer the natural habitat intensively.

## Results for the Family Carabidae

Distribution of 362 individuals belonging to 12 different species in the selected sampling areas is as follows. 234 individuals belonging to 11 species were detected from Natural habitat, 124 individuals belonging to 7 species were detected from Semi-natural habitat and 4 individuals belonging to 3 species were detected from Urbanization zone (Figure 5).

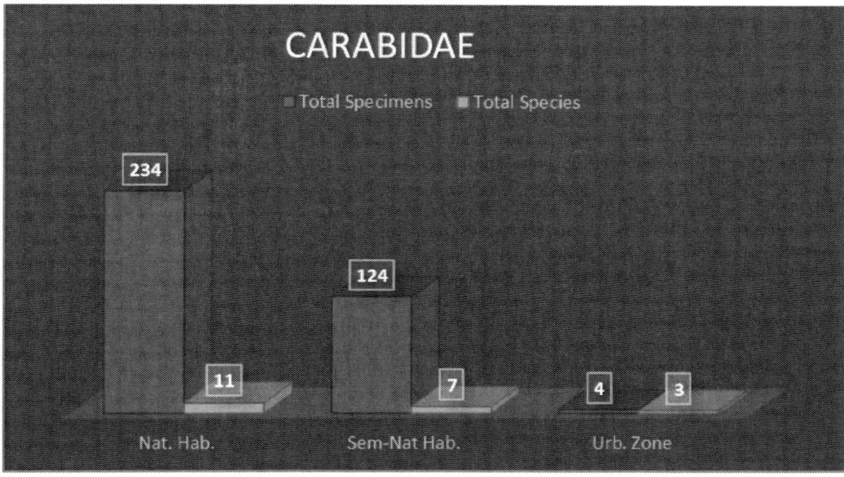


**Figure 5.** Comparison of total Carabidae species and individual numbers on study areas.

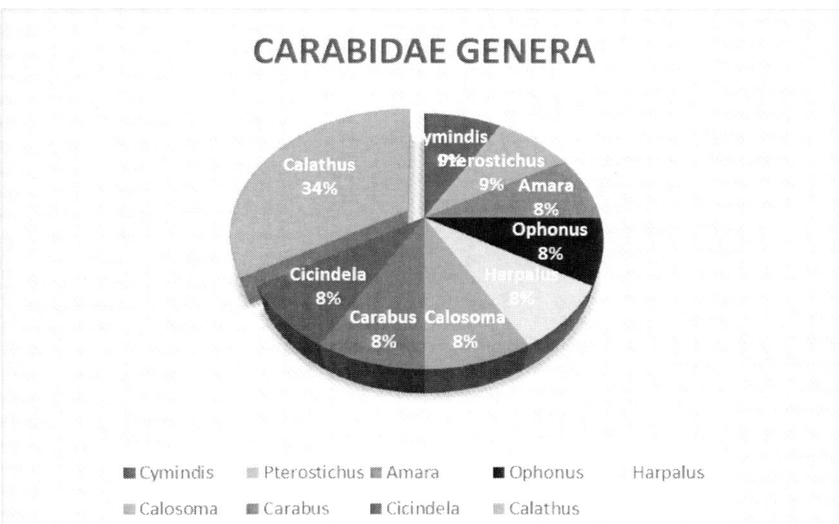


**Figure 6.** The percentages of Carabidae genera on all study areas.

The species distribution by genera is given in Figure 6. Calathus (16%), Cicindela (9%), and Carabus (9%) are the genera of Carabidae with the highest number of species in the areas, respectively. The distribution of the detected species in the areas are given in Figure 7; Figure 8 and Figure 9, orderly.

The individual numbers and percentages of the 3 species in the Urbanization zone area show that *Cicindela (Cicindela) campestris* Linne, 1758 is the most abundant species with a rate of 50%, followed equally by *Calathus (Calathus) fuscipes* (Goeze, 1777) with 25% and *Calathus (Neocalathus) cinctus* Motschulsky, 1850 with 25% (Figure 7).

The number of individuals and percentages of the 7 species in the Semi-natural habitat area show that *Calathus (Calathus) fuscipes* (Goeze, 1777) is the most common species with 64%, followed by *Carabus (Pachystus) graecus* Dejean, 1826 with 10% and *Calathus (Neocalathus) melanocephalus* (Linne, 1758) with 9% (Figure 8).

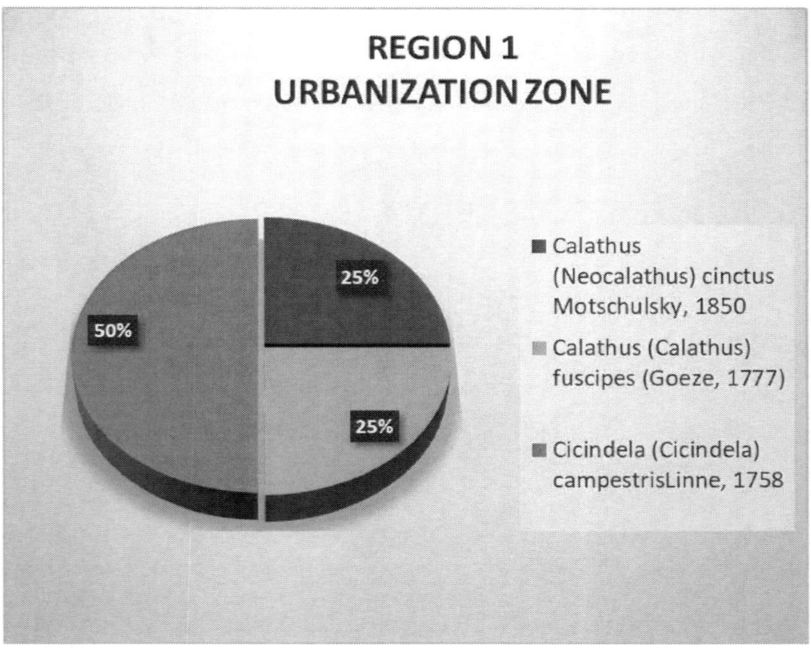


**Figure 7.** Carabidae species and percentages on Urbanization zone.

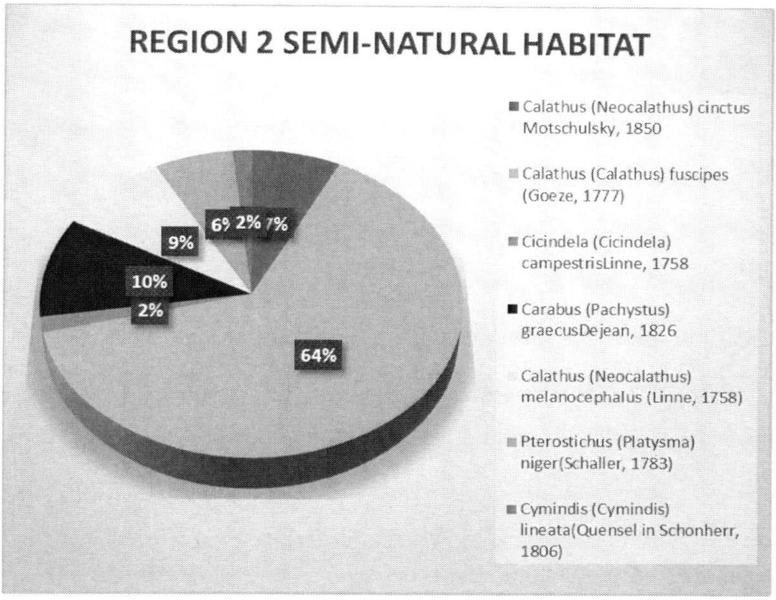


**Figure 8.** Carabidae species and percentages on Semi-Natural habitat.

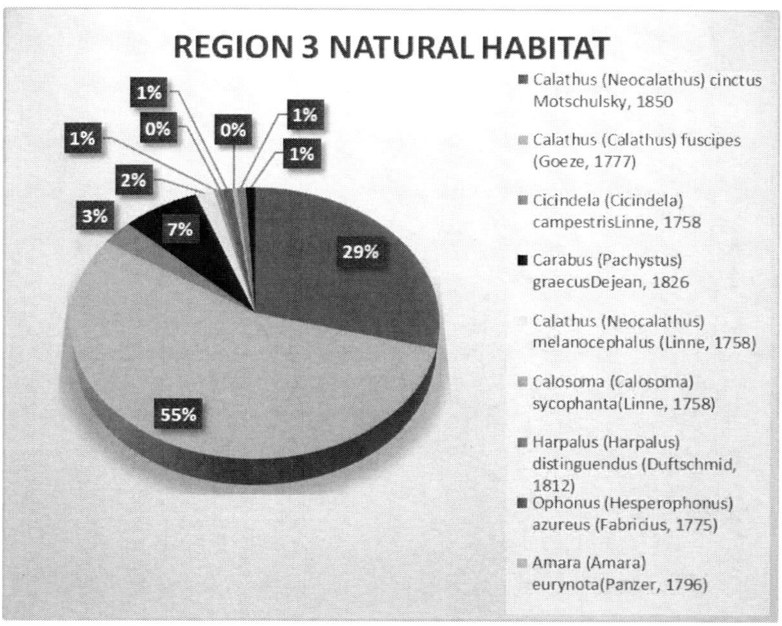


**Figure 9.** Carabidae species and percentages on Natural habitat.

The number of individuals and percentages of 11 species in the Natural habitat area show that Calathus (Calathus) fuscipes (Goeze, 1777) is the most common species with 55%, followed by Calathus (Neocalathus) cinctus Motschulsky, 1850 with 29%, and Carabus (Pachystus) graecus Dejean, 1826 with 7% (Figure 9).

## Results for the Family Scarabaeidae

The distribution of 28 individuals belonging to 7 different species in sampling areas is as follows. 23 individuals belonging to 7 species were detected from Natural habitat, 4 individuals belonging to 1 species were detected from Semi-natural habitat and 1 individual belonging to 1 species was detected from the Urbanization zone (Figure 10).

The distribution of the detected species according to genera is given in Figure 11. While Oxythyrea and Protaetia genera received 29%, Holochelus, Netocia and Cetonia genera received 14%.

The distribution and percentages of the Scarabaeidae individuals collected from the Natural Habitat are given in Figure 12. The number of individuals and percentages of the 7 species in the natural habitat area shows that *Protaetia (Netocia) cuprea* (Fabricius, 1775) is the most common species with 35%, followed by *Holochelus (Miltotrogus) sp.* Reitter, 1902 with 31%, and *Oxythyrea cinctella ab.natalia* (Olsoufieff, 1916) with 13%.

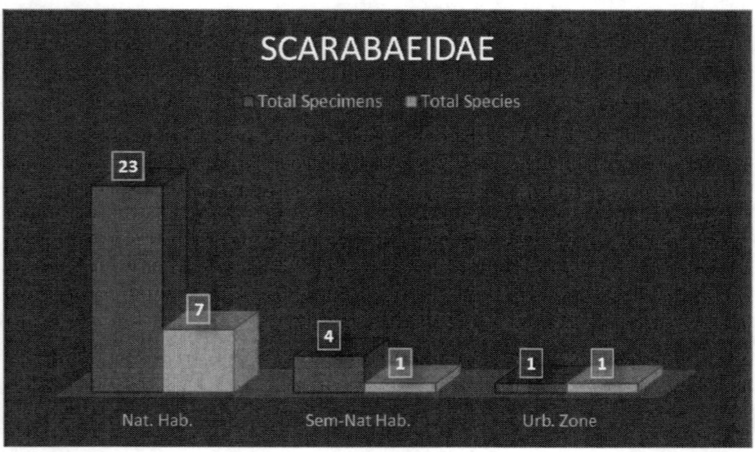


**Figure 10.** Comparison of total Scarabaeidae species and individual numbers on study areas.

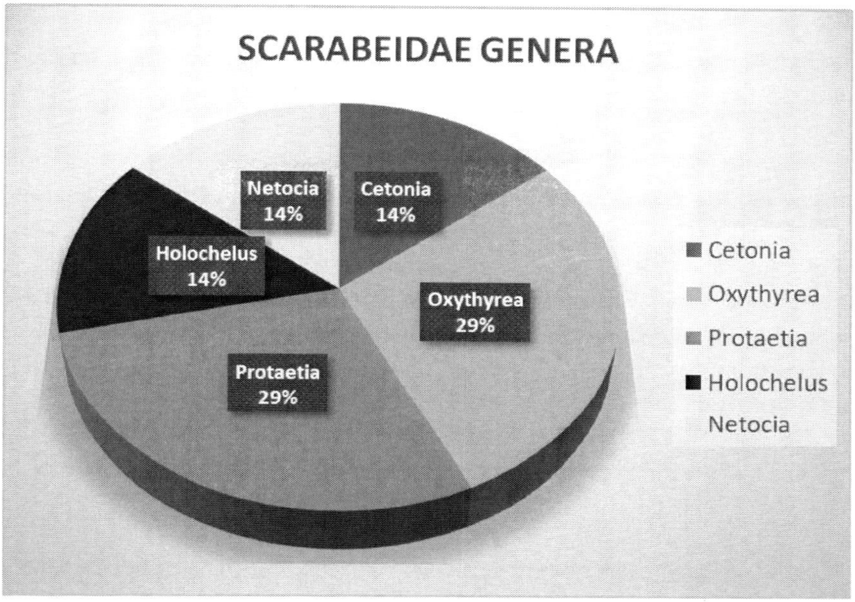


**Figure 11.**The percentages of Scarabaeidae genera on all study areas.

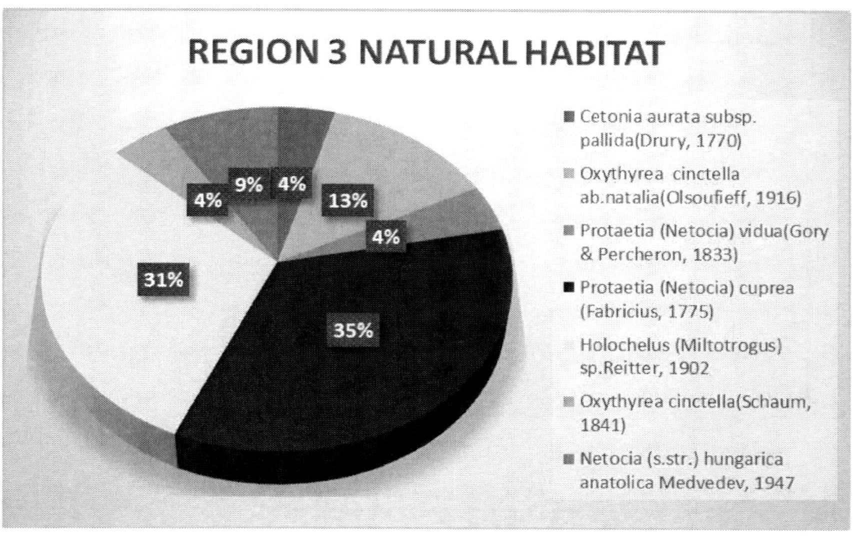


**Figure 12.** Scarabaeidae species and percentages on Natural habitat.

Only *Protaetia (Netocia) cuprea* (Fabricius, 1775) was found in the Semi-natural habitat and only 4 individuals of this species were obtained during the study. Again, only one individual of *Oxythyrea cinctella* (Schaum, 1841) was found in the urbanization zone during the study period.

*Calathus (Calathus) fuscipes* (Goeze, 1777) was the most common species with 209 individuals. This species is observed in all sampling areas even if it has a low number of individuals. Examination of the number of individuals and species of the specimens detected in both families shows that the species intensively prefers natural habitats. Similarity values between habitats were calculated based on Carabidae and Scarabaeidae individuals and species numbers. Dendograms based on the similarity index results are given in Figure 13 and Figure 14.

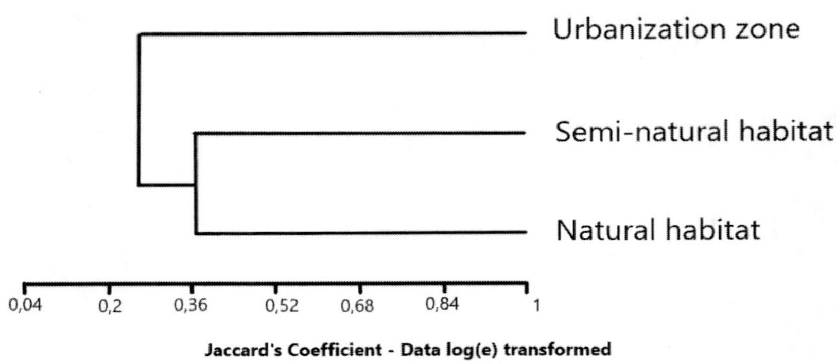


**Figure 13.** Jaccard's Similarity Dendogram.

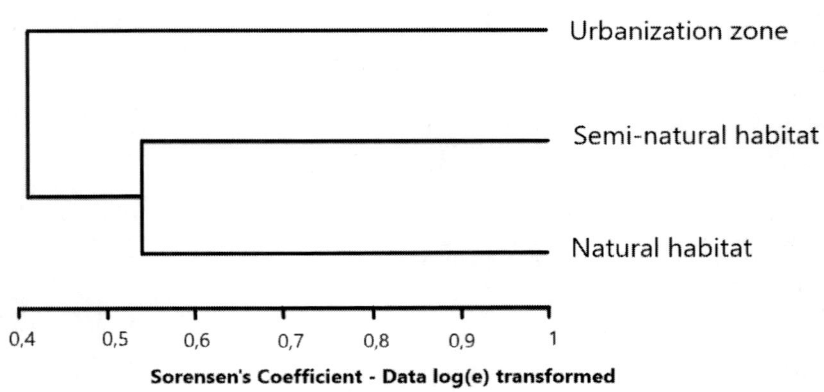


**Figure 14.** Sorensen's Similarty Dendogram.

Both in Sorensen's and Jaccard's similarity index results show that the areas are clustered in two branches. Semi-Natural Habitat and Natural Habitat were found to be the most similar areas with a Jaccard's index of 0.368 and Sorensen's index of 0.538, respectively.

Simpson and Shannon-Wiener diversity index results are shown in Table-2 and Table-3.

**Table 2.** Simpson's diversity index results

| Sample | Index | Evenness | Num.Spec. |
|---|---|---|---|
| Region 1-Urbanization zone | 0, 720 | 0,960 | 4,000 |
| Region 2-Semi natural habitat | 0,591 | 0,676 | 8,000 |
| Region 3-Natural habitat | 0, 671 | 0,710 | 18,000 |

**Table 3.** Shannon-Wiener diversity index results

| Sample | Index | Evenness | Num.Spec. |
|---|---|---|---|
| Region 1-Urbanization zone | 1,332 | 0,961 | 4,000 |
| Region 2-Semi natural habitat | 1,339 | 0,644 | 8,000 |
| Region 3-Natural habitat | 1,579 | 0,546 | 18,000 |

According to the diversity indices, although the urbanization area has the lowest number of species, it was determined as the area with the highest diversity due to the balanced distribution of the number of individuals of the species. When the areas ranked in terms of the number of species are evaluated in terms of diversity, the abundance and dominance of the species in the environment also come to the fore and affect the results significantly.

## Discussion

Comparing our study with studies investigating the effects on insect groups and their habitats; Alaruikka et al., (2002) compared how Carabidae and ground spiders were affected by urbanization and human pressure in their study in semi-urbanized forests of southern Finland. Samples were collected from 3 different zones: natural, semi-natural and urbanized. They noted that the total abundance and richness of both taxa were not significantly affected by urbanization, but there were areas where some species showed specialisation along the urbanization gradient.

In the study conducted by Rickman and Connor in 2003, the effect of urbanization on three different butterfly species feeding on Quercus agrifolia was investigated and the abundance and richness of the species in 5 different habitats were measured by different mathematical methods. It was determined that the abundance of two species increased with urbanization and one species decreased.

Venn et al., (2003) compared Carabidae species abundance, richness and diversity in areas of low human disturbance in boreal forests. It was observed that in the semi-natural environment, despite a large decrease in species richness compared to the natural area, there was not a large decrease in diversity and abundance. The species were interpreted according to their habitat specialisation, and it was concluded that Carabids are a good scale in the face of habitat degradation and pressure.

In 2005, Zanette et al., measured the effects of urbanization on bumblebees and honeybees in Southeast Brazil. Except for a few widely distributed species, it was determined that the effect of urbanization on both bumblebees and honeybees was observed significantly and especially negatively affected their nesting habits. It was concluded that for a sustainable environment, it is necessary to take precautions.

Magura et al., in 2008 investigated how Carabidae species' diversity is numerically affected and how the intensity of species activity varies in a region called semi-natural and a provincial region not exposed to human impact. In the study, environmental variables of the habitats where the species were found were measured and associated with the species. As a result of the study, it was concluded that species with a wide distribution in rural areas decreased in human-dominated areas, some species did not show a great change in both areas, and some species were surprisingly abundant in areas with high human activity.

## Conclusion

Habitat loss or habitat fragmentation as a result of urbanization generally leads to the downsizing of populations living in natural areas, the fragmentation of populations and thus of their distribution, and the decrease in species richness and diversity in habitats. To have a sustainable and healthy environment, it is necessary to recognise the living environment we are in, minimise intervention in natural systems and help their functioning. Our country has a unique biodiversity. Thanks to its climate zones and habitats, it hosts high species diversity and richness. Due to the increasing rate of urbanization and human pressure on natural habitats, a significant part of this diversity is disappearing before it can be discovered or is condemned to live in small habitats.

In the 3 study sites (Natural habitat; Semi-Natural habitat; Urbanization zone), which include Eskişehir Osmangazi University Meşelik Campus and Meşelik Forest, where human impact is gradually observed, it is seen that there is a very sharp decrease in both species and individual numbers of both Carabidae and Scarabaeidae species from the region where the urbanization effect starts. When the site similarities are evaluated, the Natural habitat, where no human impact is observed, and the Semi-Natural habitat, which is exposed to a small amount of human impact due to factors such as jogging path etc., are seen as the most similar areas. The Campus area, where the effect of urbanization is intensely observed, was grouped separately from them due to the low number of species and individuals. This study proves the negative effect of increasing human impact on animal groups. The sustainability of natural areas is very important both for the creatures in natural life and for us humans. Interfering with nature as little as possible should be essential for the continuity of humanity and the environment.

## Acknowledgment

We would like to thank to Türkiye Research Project Support Programme for Undergraduate Student 2209-A (TUBITAK) for supporting the study and Geraint Evans (Brussels, Belgium) for checking the English language of the text.

## References

Acar, C., Acar, H. and Eroğlu, E. 2007. Evaluation of ornamental plant resources to urban biodiversity and cultural changing: A case study of residential landscapes in Trabzon city (Turkey). *Building and Environment,* 42: 218–229.

Alaruikka, D., Kotze, D. J. Matveinen K. and Niemela J. 2002. Carabid beetle and spider assemblages along a forested urban–rural gradient in southern Finland. *Journal of Insect Conservation,* 6: 195–206.

Atmış, E., Özden, S. and Lise, W. 2007. Urbanization pressures on the natural forests in Turkey: An overview. *Urban Forestry and Urban Greening,* 6: 83–92.

Avgın, S.S. 2006. *Studies on the fauna and taxonomy of Carabidae (Coleoptera) in Kahramanmaraş province and its surroundings*, Çukurova University, Institute of Science, PhD. Thesis, 379 p.

Cano, E. B., and Schuster J.C., 2006. *Beetles As Indicators For Forest Conservation In Central America*. UNESCO Encyclopedia of Life Support Systems.

Doygun, H. and Alphan, H. 2006. Monitoring urbanization of İskenderun, Turkey, and its Negative İmplications. *Environmental Monitoring and Assessment*. 114:145-155.

Fidan, E. C., Şirin, Ü., Kısım, E., Destire, C. and Çalışkan, H. 2014. Contribution to the knowledge of Carabidae fauna from Eskişehir Osmangazi University Meşelik Campus. *Munis Entomology & Zoology*, 9 (1): 258-265.

Fidan, E. C., and Şirin, U. 2016. The changes related with altitudinal gradient and seasonal variation in the species composition of Carabidae (Coleoptera) in Türkmen Mountain (Eskişehir, Türkiye). *Ekoloji*, 25(98): 17-24.

Hurka, K. 1996. *Carabidae of the Czech and Slovak Republics*, Kabourek Publisher, 565 p.

Jarvis, P. J. 2000. *Ecological Principles and Environmental Issues*. Pearson Education Limited, England, 303 p.

Kovach, W.L. 1998. *MVSP - A Multivariate Statistical Package for Windows, ver. 3.0.* Kovach Computing Services, Pentraeth, Wales, U.K.

Krebs, C. J. 1994. *Ecology: The Experimental Analysis of Distribution and Abundance*, Harper Collins, New York, 801 p.

Krebs, C. J. 1999. *Ecological Methodology*. Benjamin Cummings, California, 620 p.

Löbl, I. and Smetana, A. 2004. *Subfamily Staphylininae-Catalogue of Palaearctic Coleoptera. Volume 2. Hydrophiloidea, Histeroidea, Staphylinoidea.* Apollo Books, Stenstrup 624-698.

Magura, T., Tothmeresz B., and Molnar, T. 2008. A species-level comparison of occurrence patterns in carabids along an urbanisation gradient. *Landscape and Urban Planning* 86: 134–140.

Magurran, A. E. 2004. *Measuring Biological Diversity*. Blackwell Publishing: Oxford, UK. 256 p.

McKinney, M. L. 2008. Effects of urbanization on species richness: A review of plants and animals. *Urban Ecosystem*. 11:161-176.

Morse, C. C. 2001. *The Response of First and Second Order Streams to Urban Land-use in Maine, U.S.A,* M.Sc. *Thesis,* Ecology and Environmental Science Program, The University of Maine, Orono, Maine, U.S.A.

Nahmani, J., Lavelle, P., Rossi, J.P., 2006. Does changing the taxonomical resolution alter the value of soil macroinvertebrates as bioindicators of metal pollution? *Soil Biology and Biochemistry,* 38(2): 385-396.

O'Driscoll, M., Clinton, S., Jefferson, A., Manda, A. and McMillan, S. 2010. Urbanization Effects on Watershed Hydrology and in Stream Processes in the Southern United States. *Water*, 2: 605-648.

Pearce, J. L. and Venier, L. A. 2006. The use of ground beetles (Coleoptera: Carabidae) and spiders (Araneae) as bioindicators of sustainable forest management: A review. *Ecological Indicators* 6: 780-793.

Pearsall, I. and Dunsworth G. 2003. Assesment of Carabid Beetles as Ecological Indicators. *Weyen haeuser*. 6: 1-7.

Rainio, J., and Niemela, J. 2003. Ground Beetles (Coleoptera: Carabidae) as Bioindicators. *Biodiversity and Conservation*, 12: 487-506.

Rickman, J. K. and Connor, E. F. 2003. The effect of urbanization on the quality of remnant habitats for leaf-mining Lepidoptera on *Quercus agrifolia*. *Ecography*, 26: 777-787.

Samways, M. J. 1989. Insect Conservation and landscape ecology: a case-history of bush crickets (Tettigoniidae) in southern France. *Environmental Conservation*. 16: 217-226.

Şekercioğlu, H. Ç., Anderrson, S., Akçay, E., Bilgin, R., Can, Ö. E., Semiz, G., Tavşanoğlu, Ç., Yokeş, M. B., Soyumert, A., İpekdal, K., Sağlam, İ.K., Yücel, M., and Dalfes, H. S., 2011. Turkey's globally important biodiversity in crisis, *Biological Conservation*, 12: 2752-2769.

Şenyüz, Y. 2009. *Fauna of the Turkmen Mountain Aphodiinae (Scarabaeidae, Coleoptera) Subfamily* Dumlupınar University, Institute of Science, Department of Biology. PhD Thesis. 103 p.

Shahabuddin, Hasanah, U. and Elijonnahdi 2014. Effectiveness of Dung Beetles as Bioindicators of Environmental Changes in Land-Use Gradient in Sulawesi, Indonesia. *Biotropia* Vol. 21 No. 1: 48 - 58.

Syaripuddin, K., Sing, K. and Wilson J.J. 2015. Comparison of butterflies, bats and beetles as bioindicators based on four key criteria and DNA barcodes. *Tropical Conservation Science* 8(1): 138-149.

Trotuş, E., Naie, M. 2008. Data on the Knowledge of Injurious Organisms in Hemp Crops From the Central Moldavia. *Cercetari Agronomice*, 3: 51-57.

Venn, J. S., Kotze, D. J., Niemela, J. 2003. Urbanization effects on carabid diversity in boreal forests. *European Journal of Entomology*. 100: 73-80.

Zanette, L. R. S., Martins R. P., Ribeiro S. P. 2005. Effects of urbanization on Neotropical wasp and bee assemblages in a Brazilian metropolis. *Landscape and Urban Planning* 71: 105–121.

## Chapter 9

# The Morphological Characteristics of Female *Haemonchus Contortus* Obtained from Goats in the State of Maranhão, Brazil

**Danilo Rodrigues Barros Brito[1,*], PhD**
**Joanice Francisca Machado Araújo[1], BSc**
**Livio Martins Costa Júnior[2], PhD**
**and Rafael Arturo Torres-Fajardo[2], PhD**
[1]Instituto Federal de Maranhão – IFMA, São Luís, Maranhão, Brazil
[2]Laboratório de Controle de Parasitos, Universidade Federal de Maranhão – UFMA, São Luís, Maranhão, Brazil

## Abstract

*Haemonchus contortus* is an important gastrointestinal nematode in goat and sheep farming in Brazil, because it causes great economic losses, impairs animal productivity, and in some scenarios can cause high mortality. This work aimed to describe the characteristics of the vulvar flap of *H. contortus* females infecting goats in the state of Maranhão, Brazil. Parasitological necropsies were carried out in twenty-four crossbred Boer male goats previously infected with gastrointestinal nematodes. After the animals were euthanized, the abomasum was immediately removed and washed for adult nematode recovery. 10% formalin was added to the material for further counting and identification of adult nematodes. The parasites were then separated using a stereoscopic microscope, fixed with Hoyer's medium, and subsequently identified. A total of 2,714 adult females of *H. contortus* were examined

[*] Corresponding Author's Email: danilobrito@ifma.edu.br.

In: Advances in Animal Science and Zoology. Volume 21
Editor: Owen P. Jenkins
ISBN: 979-8-89113-291-7

within which 58.6% presented a smooth type vulvar appendix, being the predominant form. Furthermore, the 26.4% and 15% of specimens had a linguiform and button type vulvar appendixes, respectively. These variations might be natural or suggest morphological adaptations according to the geographic region and the host species. These results contribute to the taxonomy and differentiation of *H. contortus* infecting goats in the state of Maranhão, Northeastern Brazil, where haemoncosis represents a threat for the goat farming.

**Keywords:** goat diseases, haemonchosis, morphological identification, vulvar flap

## Introduction

The infection with gastrointestinal nematodes (GIN) is arguably the most common health issue in productive systems where ruminants depend on natural vegetation. From the many organisms involved in this complex relationship, the genus Haemonchus, which belongs to the superfamily Trichostrongylidae and includes 12 species, is known to parasitize ruminants, either domestic or wild. Regarding domestic ruminants, H. contortus and H. placei are the most common in small ruminants and cattle, respectively (Gilleard et al., 2021). More specifically, the species Haemonchus contortus (Rudolphi, 1803) excels for its capacity to parasitize the abomasum and for its hematophagous feeding style that has been associated with an impair health, productivity, and welfare in small ruminant productive systems worldwide (Arsenopoulos et al., 2021). The clinical (ante mortem) signs associated with haemonchosis are not disease-specific but tend to guide the clinical practitioner towards a field diagnostic that must be confirmed with additional laboratory information. In this regard, and due to its blood-sucking activity, anaemia is the most reported consequence, but other signs such as submandibular edema, hair-loss, anorexia, diarrhoea, weakness, drowsiness, sub-optimal performance, emaciation, and eventually death have been reported (Starkey and Pugh, 2021; Smith and Sherman, 2023).

Morphologic identification has been used to identify ruminant haemonchosis, especially in low-income regions where the access to more accurate methods is somewhat impractical and certainly unaffordable (Zarlenga et al., 2016). Several morphological features in Haemonchus sp., especially for the livestock important species H. contortus and H. placei, have been developed to identify the causal organism of parasitic infection of ruminants within which the vulvar female anatomy and some phenotypic variations on the vulvar flap have been used for many decades as a primary basis for species and subspecies designations (Gibbons, 1979).

In the hope of deepening our understanding of the biology, morphology, and phenotypic variations of gastrointestinal nematode infections in small ruminants, the present work aimed to describe the characteristics of the vulvar flap of Haemonchus contortus females infecting goats in the state of Maranhão, Brazil.

## Materials and Methods

### Study Area

The experiment was performed in the small ruminant sector at Instituto Federal de Maranhão (IFMA), campus Maracanã, São Luiz, Maranhão, Brazil (2° 37′ 01″ S and 44° 16′ 19″ W). Climate in the urban and suburban region of São Luis is, according with the classification of Köppen (1928), hot and sub-humid presenting a mean annual temperature of 26.8 ° C and a mean annual precipitation of 2,156 mm (Climate-Data.org, 2021). All the procedures performed with animals were previously approved by the ethics and animal experimentation commission at Universidad Estadual de Maranhão under the approval number 015/2012.

### Experimental Goats

A total of 24 castrated juvenile male mixed-boer goats were used in the present experiment. These animals aged six-to-eight months and weighed 15 ± 2.5 kilograms of liveweight (LW). Animals were maintained in individual pens with *ad libitum* access to fresh water, mineral supplementation (Caprinofós®, DSM), and an integral ration that was elaborated with 35% grounded corn, 9% wheat flour, and 7% soybean flour. Animals were daily supplemented with an equivalent of 3% their LW divided into two moments (08:00 and 16:00). Finally, 1,200 grams in fresh basis of *Cynodon* spp. hay were administered to each goat.

## Gastrointestinal Nematode Infection and Parasitological Assessments

An artificial infection protocol was established for each experimental animal. Within this process 16,000 infective larvae ($L_3$) were orally administered in a single dose. The GIN $L_3$ were obtained from a contemporary experiment at IFMA culturing the faeces of adult goats carrying a natural infection. After the $L_3$ administration, samples of faeces were directly taken from the rectum of each experimental goat on a weekly basis to identify and quantify the excretion of GIN eggs and to estimate the nature of the infection. Assessment of the GIN egg excretion was performed by means of the Gordon and Whitlock technique with some modifications (Ueno and Gonçalves, 1998). Faecal cultures were performed following Roberts and O'sullivan (1950) and identification of $L_3$ was based on the descriptions of Ueno and Gonçalves (1998). On the day 28$^{th}$ after the artificial infection protocol, daily faecal samples were taken from each animal during one additional week.

## Collection of Adult *Haemonchus Contortus* Females

After 35 days of artificial infection and adult nematode establishment, goats were fasted for 24 hours and humanely euthanized. The procedures to obtain adult abomasal nematodes were performed as by Torres-Acosta et al., (2015). Briefly, abomasum, small intestine, and large intestine were identified and tightened with a double ligature in the abomasal-duodenal junction, ileo-caecal junction, and rectum, respectively. Abomasum was separated from the omasum, deposited in a recipient, and opened along its greater curvature. After this, the mucosal content was thoroughly washed using a sieve no. 400 and the remaining material was placed in a plastic container with 10% formalin until posterior analyses.

## Vulvar Morphology

Nematode females were separated from males by using a stereoscopic microscope (OPTICAM O400S, São Paulo, Brazil). Preparation of plates for structural visualization was done by fixing the specimens with Hoyer's solution. After this, all specimens were visualized for the presence of vulvar flap types following the guidelines of Lichtenfels et al., (1994) and Vicente et al., (1997). Broadly, the vulvar types were identified as: linguiform (presenting a supravulvar flap), knobbed (presenting a knob-like vulvar process) and smooth (without an evident vulvar process).

## Data Analysis

The Microsoft Office Excel® program, version 2010, was used for data processing. Data analysis and interpretation were carried out according to the descriptive method and the tabular analysis technique, using the simple rule of three to arrive at the exact result of the percentage.

## Results

The weekly follow-up of GIN excretion in the goats allowed us to confirm the presence of some eggs in faeces after the third week of sampling. Faecal culture data verified the multispecific nature of GIN infection, within which the genera *Haemonchus* sp., *Trichostrongylus* sp., and *Oesophagostomum* sp. comprised the 53.5%, 40.2%, and 6.3% of infection, respectively.

A total of 5,240 adult *H. contortus* were recovered from the abomasum of experimental goats. From this, 48.2% (n = 2,526) and 51.8% (n = 2,714) were males and females, respectively. Selected females were subsequently used and after a thorough identification of the vulvar appendix, the smooth type was present in the 58% (n = 1,574) of evaluated specimens, being the predominant form. In addition, the linguiform and knobbed types were identified in the 26% (n = 705) and 16% (n = 435) of female nematodes, respectively.

## Discussion

For the best of our knowledge, the present study reported, for the first time, the anatomical features in the vulvar flap of adult *H. contortus* females infecting goats in the state of Maranhão, Northeastern Brazil. Other country reports were previously produced in the states of Rio Grande do Norte (Oliveira Nascimento et al., 2013) and Ceará (Moreira das Neves et al., 2012), both belonging to the north-eastern region of the country and presenting a semiarid climate which differs from the hot and sub-humid conditions in our state. Goat production in the state of Maranhão represents an activity with a high social relevance since most farmers have low income and scarce access to technologies. Furthermore, the climatological conditions of the state are of epidemiological interest considering that it represents a transition area amongst the Amazonian Forest, the northeastern region, and the semiarid region of the country.

The present study reported a predominance of the smooth type vulvar flat. This observation coincides with Mejía-García and Díaz-Benítez (1987) and Moreira das Neves et al., (2012), but is at odds with other reports in which either the linguiform (Jacquiet et al., 1995; Thomas et al., 2007; Kumsa et al., 2008; Gharamah et al., 2011; Moreira das Neves et al., 2012; Akkari et al., 2013; Oliveira Nascimento et al., 2013) or the knobbed (Rahmad and Hamid, 2007; Gharamath et al., 2011; Akkari et al., 2013; Degheidy et al., 2014; Nahar et al., 2019) types were predominant. The table 1 presents a summary of the published information reporting the structural features of the vulvar flap of female *H. contortus* infecting small ruminants worldwide. It is out of the scope of the present work to perform detailed comparisons amongst these works since they were performed in different agroecological and productive conditions, used different host species, breeds, season of the year, amongst other influential factors. Nonetheless, we want to highlight the variability and plasticity that *H. contortus* display within the revised studies. Such plasticity might be one of the characteristics allowing its widespread nature on a range of environments and small ruminant productive systems (Arsenopoulos et al., 2021).

Considering the above-exposed, the adaptive value of the vulvar flap merits a brief discussion. On the one side, it is quite normal that these morphological variations occur within a nematode population. For certain authors, notwithstanding, these variations could increase the nematode adaptability, serve as a marker of ecological adaptations, provide taxonomic relevance, allow the understanding of nematode biology, and determine the parasitic populations in specific areas (Jacquiet et al., 1995; Kumsa et al., 2008). This approach was established since Chitwood (1957) and Das and Whitlock (1960) who established three basic types of vulvar flat: linguiform, knobbed, and smooth. Upcoming works have proposed additional classification schemes. LeJambre and Whitlock (1968) classified certain sublinguiform types into A, B, C, and I. An upcoming work aimed to determine the type of vulvar processes of adult GIN infecting sheep, goats, and cattle in Tunisia identified the new 'LD' sublinguiform vulvar flap type (Akkari et al., 2013).

**Table 1.** Proportion of *Haemonchus contortus* vulvar types infecting small ruminants and reported for different agroecological conditions

| Country | n | Host | Vulvar type (%) | | | Reference |
|---|---|---|---|---|---|---|
| | | | Smooth | Linguiform | Knobbed | |
| México | 876 | Goat | **73.5** | 3.4 | 21.9 | Mejía-García and Díaz-Benitez, 1987 |
| Mauritania | 251 | Goat | 14 | **56** | 30 | Jacquiet et al., 1995 |
| Mauritania | 551 | Sheep | 16.5 | **55.5** | 28 | Jacquiet et al., 1995 |
| Malaysia | 100 | Goat | 10 | 26 | **64** | Rahmad and Hamid, 2007 |
| Malaysia | 100 | Sheep | 15 | 31 | **54** | Rahmad and Hamid, 2007 |
| Ethiopia | 448 | Goat | 29 | **43.8** | 27.2 | Thomas et al., 2007 |
| Ethiopia | 653 | Sheep | 35.4 | **37.8** | 30 | Thomas et al., 2007 |
| Ethiopia | 2386 | Goat | 27.6 | **53.8** | 18.4 | Kumsa et al., 2008 |
| Ethiopia | 3186 | Sheep | 23 | **49.9** | 28.5 | Kumsa et al., 2008 |
| Malaysia | 50 | Goat | 14 | 16 | **70** | Gharamah et al., 2011 |
| Malaysia | 50 | Sheep | 6 | **54** | 40 | Gharamah et al., 2011 |
| Brazil | 100 | Sheep | 32.1 | **55.5** | 12.4 | Moreira das Neves et al., 2012 |
| Brazil | 100 | Sheep | **49** | 41.3 | 9.8 | Moreira das Neves et al., 2012 |
| Tunisia | 645 | Goat | 6.4 | 27.7 | **65.8** | Akkari et al., 2013 |
| Tunisia | 1662 | Sheep | 21.2 | **42.5** | 36.3 | Akkari et al., 2013 |
| Brazil | 50 | Goat | 18 | **72** | 10 | Oliveira Nascimento et al., 2013 |
| Saudi Arabia | 300 | Goat | 18 | 23 | **50** | Degheidy et al., 2014 |
| Bangladesh | 520 | Goat | 20.4 | 33.5 | **46.1** | Nahar et al., 2019 |
| Brazil | 2714 | Goat | **58** | 26 | 16 | This work |

n: number of evaluated *Haemonchus contortus* females

Numbers highlighted in bold represent the most common vulvar type reported for each study.

The present work allowed us to identify and describe some anatomic features of the most parasitic and threatening gastrointestinal nematode for goat production worldwide. We also evidenced an inherent variability in the vulvar flap of nematode females infecting juvenile goats in an experimental farm situated in the hot and sub-humid northeastern region of Brazil and this could contribute to the understanding of the parasite biology and the complex relationships between these organisms, their hosts, and the environment. However, we acknowledge that further research is needed in the specific conditions of the Maranhão state to better understand the prevalence, ecology, morphology, molecular, and genetic features of gastrointestinal nematodes, especially in tropical and subtropical regions where haemonchosis represent a hurdle for the profitability of goat production.

## Conclusion

The outcomes from the current study allowed us to categorize the vulvar flap anatomy of *Haemonchus contortus* females parasitizing the abomasum of juvenile mixed-boer goats with an artificial nematode infection under the conditions of São Luis Maranhão, Brazil. The most prevalent vulvar type was smooth, followed by the linguiform, and knobbed. It would be interesting to explore whether this pattern is maintained in other small ruminant regions of the state and whether the variability in the vulvar anatomy provides an ecological advantage for these organisms.

## References

Akkari H, Gharbi M, Awadi S, Mohamed AD, Kumsa B. New sublinguiform vulvar flap of Haemonchus species in naturally infected domestic ruminants in Béja Abattoir, North Tunisia. *Veterinarski Arhiv* (2013) [Veterinary Archive] 83(3), 281-291.

Arsenopoulos K V, Fthenakis G C, Katasarou E I, Papadopoulos E. Haemonchosis: A challenging parasitic infection of sheep and goats. *Animals* (2021) 11, 363.

CLIMATE-DATA.ORG. Weather data for world cities. Available in: https://pt.climate-data.org/america-do-sul/brasil/maranhao-206/. Consulted in: August 11. 2023.

Chitwood MB. Intraspecific variation in parasitic nematodes. *Systematic Zoology* (1957) 6, 19-23.

Das K M, Whitlock J H. Subspeciation in Haemonchus contortus (Rudolphi, 1803) Nematoda, Trichostrongyloidae. *Cornell Veterinary* (1960) 50, 182-197.

Degheidy N S, Al-Malki J S, Al-Olmari F I. Some epidemiological estudies of caprin Haemonchosis in Taif, Saudi Arabia. *International Conference on Advances in Agricultural, Biological & Environmental Sciences (AABES-2014)* (2014) Oct 15-16 Dubai (UAE).

Gibbons L M. Revision of the genus Haemonchus Cobb, 1988 (Nematoda: Trichostrongylidae). *Systematic Parasitology* (1979) 1(1), 3-24.

Gilleard J S, Kotze A C, Leathwick D, Nisbet A J, McNeilly T N, Besier B. A journey through 50 years of research relevant to the control of gastrointestinal nematodes in ruminant livestock and thoughts on future directions. *International Journal of Parasitology* (2021) 51:1133-1151.

Gharamah A A, Rahman W A, Nor S A M. Phenotypic differences of Haemonchus contortus from sheep and goats in the states of Perak and Kelantan, peninsular Malaysia. *Acta Parasitologica* (2011) 56(4), 412-417.

Jacquiet P, Humbert J F, Comes A M, Cabaret J, Thiam A, Cheikh D. Ecological, morphological and genetic characterization of sympatric Hemonchus spp. parasites od domestic ruminants in Mauritania. *Parasitology* (1995) 110, 483-492.

Köppen W, Geiger R. Klimate der Erde. Gotha: *Verlag Justus Perthes.* Wall-map 150 x 200cm. 1928.

Kumsa B, Tolera A, Abebe R. Vulvar morphology and sympatry of Haemonchus species in naturally infected sheep and goats of Ogaden region, eastern Ethiopia. *Veterinarsky Arhiv* (2008) [Veterinary Archive] 78(4), 331-342.

LeJambre L F, Whitlock J H. Seasonal fluctuation in linguiform morphs of Haemonchus contortus cayugensis (1968) *Journal of Parasitology* 54, 827-830.

Lichtenfels JR, Pilitt PA, Hoberg EP. New Morphological Characters for Identifying Individual Specimens of Haemonchus spp. (Nematoda: Trichostrongyloidea) and a Key to Species in Ruminants of North America. *Journal of Parasitology* (1994) 80, 107–119.

Mejía-García R A, Díaz Benitez A. Variación morfológica del apéndice vulvar de Haemonchus contortus y H. similis en bovinos y caprinos. *Revista Iberoamericana de Parasitología* (1987) 47(4), 365-375. [Morphological variation of the vulvar appendage of Haemonchus contortus and H. similis in cattle and goats. *Ibero-American Journal of Parasitology*]

Moreira das Neves M R, Vieira L S, Morais dos Santos C, Benvenuti C L, Carmo Navarro A M, Lemos de Brito R L, Zaros L G. Identificacao das espécies de nematoides gastrintestinais em ovinos 1/2 sangue Santa Inés. [Identification of gastrintestinae nematoid species in sheep 1/2 Santa Inés blood.] Proceedings of VII Congreso Nordestino de Producao Animal. (2012). Maceió, Alagoas, Brazil.

Nahar L, Sarder M J U, Rahman M, Begum M I A, Rahman S. Vulvar flap morphology of Haemonchus contortus in naturally infected slaughtered goats in Northern area of Bangladesh. *Bangladesh Journal of Veterinary Medicine* (2019) 17(2), 125-131.

Oliveira Nascimento J, Soares Pereira J, Souza Fonseca Z A A, Costa Coelho W A, Bessa N A, Mendes Ahid S M. Aspectos morfométricos de Haemonchus contortus (Rudolphi, 1803) procedentes de caprinos (Capra hircus) da região semiárida do Rio Grande do Norte, Brasil. Arquivos do Instituto Biologico de São Paulo (2013) 80(4), 447-449. [Morphometric aspects of Haemonchus contortus (Rudolphi, 1803) from goats (Capra hircus) from the semi-arid region of Rio Grande do Norte, Brazil. Archives of the Biological Institute of São Paulo].

Roberts F H S, O'sullivan J P. Methods for egg counts and larval cultures for strongyles infesting the gastrointestinal tract of cattle. *Australian Journal of Agricultural Research* (1950) 1, 99-102.

Smith, Mary C., and Sherman, David M. 2023. *Goat Medicine*. Third edition. Wiley Blackwell.

Starkey, Lindsay A., and Pugh, David G. 2021. Internal parasites of sheep, goats, and cervids. In: Pugh D G, Baird AN, Edmonson M, Passler T (Eds). *Sheep, goat, and cervid medicine*. Elsevier. Third edition. 97-117

Thomas N, Teshale S, Kumsa B. Abomasal nematodes of sheep and goats slaughtered in Awassa (Ethiopia): species composition, prevalence and vulvar morphology. *Helminthologia* (2007) 44(2), 70-75

Torres Acosta J F J, Vargas Magaña J J, Chan Pérez J I, Aguilar Caballero A, Alonso Díaz M, Ojeda Robertos NF. Recuperación de helmintos a la necropsia. In: Rodríguez Vivas Roger Iván (Ed). *Técnicas para el diagnóstico de parásitos con importancia en salud pública y veterinaria*. 158-199.

Ueno, H., and Goncalves, P C. 1998. *Manual para diagnóstico das helmintoses de ruminantes*. [Manual for diagnosing ruminant helminthiasis]. Fourth edition. Japan International Cooperation Agency. Tokyo, Japan.

Vicente J J, Rodrigues H O, Gomes D C, Pinto R M. Nematóides do Brasil. Parte V: nematóides de mamíferos. Brazillan nematodes. Part V: nematodes of mammals. *Revista Brasileira de Zoologia* (1997) 14 supl.1.

Zarlenga DS, Hoberg EP, Tuo W. The Identification of Haemonchus Species and Diagnosis of Haemonchosis. *Advances in parasitology* (2016) 93, 145–180.

## Biographical Sketch

### Danilo Rodrigues Barros Brito

**Affiliation:** Federal Institute of Maranhão Education: Doctor in Animal Science

**Business Address:** Avenida dos Curiós S/N Vila Esperança, São Luís, Maranhão, Brasil. Research and Professional Experience: Animal Health; Veterinary Parasitology.

**Professional Appointments:** Member of the Maranhão Academy of Science

**Honors:** 1st overall in the Planeta Play category, at the Scientific Initiation Seminar (SEMIC) - IFMA Universe 2022, IFMA.

## Publications from the Last 3 Years

1) Evaluation of a combination of Citrus aurantium var. Dulcis essential oil and albendazolefor the treatment of sheep gastrointestinal nematodes. *Veterinary Parasitology*, v. 318, p. 109929, 2023.
2) Resistance of gastrointestinal nematoids of goats and sheep to the antihelmintics levamisol, ivermectin and albendazole. *Ciência Animal Brasileira*, v. 24, p. 1, 2023.
3) Biochemical profile of castrated and uncastrated male goats supplemented with vitamin Eor not. *Revista Brasileira De Saúde E Produção Animal (Ufba)*, v. 23, p. 01-13, 2022.
4) Seroprevalence of Neospora caninum and risk factors associated with infection in water buffaloes (Bubalus bubalis) from Maranhão State, Brazil. *Veterinary Parasitology: Regional Studies And Reports*, v. 27, p. 100661, 2022.
5) Nemabiome metabarcoding reveals differences between gastrointestinal nematode speciesinfecting co-grazed sheep and goats. *Veterinary Parasitology*, v. 289, p. 109339, 2021.
6) Economic analysis of the use of wet beer residue in sheep diets. Boletim de indústria animal(Online), v. 78, p. 1-15, 2021.
7) Clinical parameters of goats infected with gastrointestinal nematodes and treated with condensed tannin. *Semina. Ciências Agrárias (Online),* v. 41, p. 517-530, 2020.
8) Ocorrência de doenças infecciosas e parasitárias em caprinos e ovinos da região metropolitana de São Luís, Estado do Maranhão, Brasil. *Research, Society And Development*, v. 9, p. 1-20, 2020.

## Commentary

# Don't Liberate the Animals; Let them Resist!

**Michael Allen**
East Tennessee State University, Johnson City, Tennessee, USA

### Abstract

By contrast with animal liberation, the novel concept of animal resistance has gained considerable currency in the present discourse of animal rights. To what extent does animal resistance represent a genuine alternative to the more established paradigm of animal liberation? I argue that animal resistance is less an alternative to the liberationist paradigm than a way for liberationists to reconnect with earlier abolitionist campaigns analogizing animal and human slavery. Indeed, analogizing animal and human resistance combines empathy for all suffering animals with higher order human capacities for corporate, or group, level thinking. This helps motivate moral judgments concerning the duty to liberate all animals suffering cruelty and oppression.

**Keywords:** animal liberation, animal resistance, analogy, empathy, suffering

Recent animal rights discourse has shifted from a paradigm of liberation to one of resistance. What motivates this shift and what is at stake in it? One rather obvious answer to the first part of this question is an increasing awareness among researchers and public alike of the extent to which captive domestic animals resist, escaping captivity as individuals or groups, simply refusing to cooperate with their captors, as well as fighting back against abusive captors (Colling 2020; Allen and von Essen 2018; Carter and Charles, 2013; Hribal 2011). Likewise, wild animals often resist hunters by adopting

In: Advances in Animal Science and Zoology. Volume 21
Editor: Owen P. Jenkins
ISBN: 979-8-89113-291-7

wily counter strategies (Allen and von Essen 2020). In addition, public awareness of animal resistance is fostered by sundry social media postings (Allen and von Essen 2018), drone and trap camera footage of animals seemingly resisting captors or hunters (von Essen, et al., 2022). As for the second part of the question about what is at stake, this concerns the status of the liberationist paradigm. Is it coherent and fruitful for humans to continue thinking of themselves as the liberators of animals if they themselves can resist their human oppressors? Hence, the new paradigm of animal resistance might be: don't liberate the animals; let them resist and liberate themselves! Humans should get over themselves thinking they are the liberators and 'savors' of the animals. They should respect the animals as properly agents of resistance and liberation by their own lights. Or should they be rather more cautious about assigning animals such a responsibility for their own liberation?

To what extent, if at all, does a resistance paradigm offer a genuine alternative to liberationism? I argue, while no such alternative, animal resistance invites revaluation of the twentieth century animal liberation movement reconnecting it – in some key respects -- with its nineteenth century legacy.

In his seminal *Animal Liberation* (1975), Peter Singer repudiates the sentimentalism about animals characterizing the earlier animal rights movement whose. Nineteenth century campaigners, like Henry Burgh, had sought an empathetic basis for the rights of suffering animals by appeal to the analogy of human and animal slaves (see my Chapter 2 in Patterson et al., 2022). By contrast, Singer focuses on rational moral judgment as opposed to feeling and empathy. He contends that if a sentient human or nonhuman is suffering, then there is "no moral justification for refusing to take that suffering into consideration" (Singer, 1975, 8). This entails making a moral choice to minimize the suffering of others, including the nonhuman. Such a choice can be made only by moral agents whose capacities extend beyond mere sentience – as the capacity for suffering – to higher order cognitions. Moral agents refusing to make such a choice are guilty of the fault of speciesism, making the "boundary of one's own species" (Singer 1975, p. 19) the boundary of moral consideration.

To be sure, Singer does not explicitly say that only human are moral agents capable of making the choice not to be a speciesist by giving equal moral consideration to the interests of all sentient beings in the minimization of their suffering. Here Singer clearly places the onus on human agents possessing moral capacities to make this choice. It is thus for them to *liberate* nonhuman species from the sufferings inflicted on them in the meat production industry, and so forth. Among other strategies for liberating animals, this means humans putting theory into practice by choosing to become vegetarians.

How then does animal resistance shift the paradigm of liberationism? It does not entail any similar demand on nonhuman species that they should refuse to be speciesist and change their dietary habits. For example, wild animals resisting human hunters by way of cunning and wiles are themselves often hunters of other wild animals. As themselves hunters, they are therefore as 'speciesist' as those humans who hunt them. While their capacities to resist their hunters may well entail agential capacities beyond the mere capacity to suffer (von Essen, et al., 2022), these need not be *universalizing moral capacities* to consider equally the interests of all others. In other words, resistance by animals entails more than mere sentience, but it need not extend to those higher order cognitive capacities stressed by Singer (1975) as necessary to calculate rationally the minimization of sufferings across species lines. This raises a perplexing question about the status of animals as *political* agents, if not universalizing moral agents. 'Resistance' is a political concept appealing less to the idea sentience, or the capacity to suffer, than *oppression*. However, 'oppression' is a corporate concept. It signifies a "social injustice" that "happens to only as the member of a group" (Cudd 2005, p. 21).

On the one hand, it entails an oppressor group, for example, the human group of nonhuman-meat consumers. On the other hand, it entails a corresponding oppressed group, the nonhuman group of industrial factory farm animals raised for consumption by the oppressor human group. That said, the concept of oppression also entails "corporate level *beliefs*" (Allen and von Essen 2018, 16). For the human oppressor group to *resist* the social injustice of oppressive factory farming, its members must entertain the group or corporate-level belief that factory farms are unjust. They subject the oppressed group of nonhumans to conditions violating their interests in not suffering, along with interests in freely exercising whatever agential capacities they might possess.

Such corporate level beliefs are the basis for the human group of oppressors exercising political agency. Acting upon such beliefs, they may resist oppression by organizing an 'animal rights' movement pursuing appropriate strategies – organizing boycotts of factory farm products, disseminating information about how to become a vegetarian (Singer 1975, Ch. 4) – to liberate the oppressed group of nonhumans. However, it is anything but clear that nonhumans farmed for consumption by humans have equivalent capacities for entertaining corporate level beliefs (Carter and Charles, 2013), that is, beliefs about themselves as members of an oppressed group, let alone beliefs about humans as an oppressor group (Fotopoulos and Sargis 2005; Staudenmaier 2004). Does this mean that farmed animals cannot be political resistors? Such animals may 'run away' or fight their way out of factory farm facilities (Colling 2020; Allen and von Essen 2018), but these are likely just individual volitional acts falling short of any *political* mode of agency that might properly be called 'resistance.' If they were capable of political resistance, then the oppressed factory farm animals would exercise corporate agency by organizing unions and adopting appropriate corporate-level strategies to resist the group of human oppressors, liberating themselves?

Likewise, hunted animals would organize as species to resist human hunters, or perhaps even organize cross species coalition of the hunted, and so on. While there is some evidence of group coordination among nonhuman animals, such as group coordinated escapes from human facilities (Colling 2020; Allen and von Essen 2018), these do not appear to rise to the level of sophisticated corporate thinking. Indeed, they do not appear to rise to the level of corporate thinking about the moral duties of membership in an oppressor or oppressed group (Vasanthakumar 2019).

Are animals therefore disqualified as resistors given their rather dubious status as corporate level thinkers? If they are to be cast as resistors, then obviously the demand for them to be higher-order corporate level thinkers must be relaxed. How might this be accomplished. One way would be to appeal to McGary's (1989) notion of an *appropriate description* of resistance to oppression. Provided an observer may describe the intentional, or volitional, act of a member of an oppressed group as causing an overall reduction of group oppression, then this observer appropriately describes it as an act of resistance. Hence, the renegade 'Tamworth Two' pigs, Butch and Sundance, performed an act of resistance, appropriately described. Their escape on transit from farm to slaughterhouse caused a media sensation leading to public criticism of the plight of slaughterhouse animals as an oppressed group (Allen and von Essen 2018). The virtue of this appeal to appropriate description is that it attributes corporate level thinking not to the individual animal performing the act of resistance, but rather to the observer and the addressees of the latter's description of the act.

Another way to relax the demand for corporate level thinking by animals would be an appeal to analogical natural law (Milbank 2021; also see Finnis 2020). Here the demand for such a univocal criterion for resistance is relaxed by appealing instead to multiple analogical similarities and differences between species and natures. Applied to both animals and humans, concepts such as rights, liberation or, for that matter, *resistance* need not match in every respect. It is sufficient for 'animal resistance' that the analogy should prove coherent and fruitful for considering the various interests, ends, and good of all species (see my chapter in this volume).

How, though, might appropriate description or analogy as resistors of their own oppression prove coherent and fruitful? Does it provide an alternative to the liberationist paradigm? At most, appeal to the novel concept of animal resistance would permit those human agents recognizing the plight of animals to describe them as acting in ways analogous to how they would act if similarly situated as members of an oppressed group. That is, they would employ whatever powers of agency they have at their disposal to resist their oppressors – escaping, fighting back however they can. Nevertheless, this quickly turns to disanalogy. Despite being capable of many ad hoc acts of resistance to their human oppressors, animals remain wholly dependent of humans to organize and mobilize an effective *movement* to resist the group oppression they suffer at the hands of humans. If this is the case, then animal resistance is no alternative paradigm. Animals continue to depend on humans for their liberation from oppression by humans.

Nevertheless, animal resistance challenges coherently and fruitfully Singer's repudiation of the sentimentalist, or empathy-based, approach of earlier animal campaigners like Burgh. Depending on the methods of appropriate description and analogy, it reconnects with a key aspect of the earlier abolitionist and liberationist movements. This is the early campaigners' appeal not to rational moral judgment as much as analogical continuity between the sufferings of human and animal slaves. Granting that no analogy is exact, Burgh's description of the sufferings of live sea Turtles shipped from the Carolinas to New York city as analogous to the sufferings endured by human slaves on the transatlantic 'middle passage' invoked a shared experience of suffering across species lines. It engaged the imagination and emotions of many New Yorkers whom Burgh mobilized successfully in campaigns reducing – if not outright abolishing – those oppressive cruelties endured not only by Turtles, but also 'fighting dogs,' and 'shooting pigeons' (Freeburg 2020).

Likewise depending on inexact cross-species comparisons, analogical descriptions of animal resistors function in much the same way. Analogically characterizing the Tamworth Two as the heroes Butch and Sundance engaged the public imagination, facilitating empathetic identifications with otherwise unseen and ignored slaughterhouse animals. Here it might be objected that empathy alone cannot sustain an abolition and liberation movement. That is true. However, is it equally true that the public is often unmoved to action, even fatigued (see my Chapter 7 in Patterson et al., 2022), by dry, censorious moral judgments of speciesism delivered by liberationists. Abolition and liberation movements may benefit from adopting a more multi-faceted approach, combining the claims of reason and sentiment.

Consistent with earlier campaigners analogizing animal slavery, analogizing animal resistance appeals to human empathetic capacities to identify with the members of other species, their sufferings, along with their interests in being liberated from cruelty and oppression. It combines this appeal to empathy with the higher order human capacities for corporate, or group, level thinking necessary to formulate rational moral judgments about their duty to liberate all animals suffering oppression.

## Disclaimer

None.

## References

Allen M. Origins of the Movement: Empathy and Emancipation, *Rethinking the American Animal Rights Movement*, Patterson-Kane E, Allen M, Edie, J. (Routledge, 2022).

Allen, M. Legacy: Overcoming Confusion and Fatigue. *Rethinking the American Animal Rights Movement*, Patterson-Kane E, Allen M, Edie, J. (Routledge, 2022).

Allen M, von Essen E. "Animal Resistors: On the Right of Resistance and Human Duties of Non-Return and Abolition." *Journal for Critical Animal Studies* (2018) 15:6: 3-28.

Carter B, Charles N. Animals, Agency, and Resistance. *Journal for the Theory of Social Behavior* (2013) 43(3): 322-340.

Colling, S. *Animal Resistance in the Global Capitalist Era.* (University of Michigan Press, 2020).

Cudd, A. How to Explain Oppression: Criteria of Adequacy for Normative Explanatory Theories. *Philosophy of the Social Sciences* (2005) 35(1): 20-49.

Finnis, J. The Nature of Law. *Cambridge Companion to the Philosophy of Law*, ed. Tasioulas J. (Cambridge University Press, 2020).

Freeburg, E. *A Traitor to His Species: Henry Burgh and the Birth of the Animal Rights Movement*. (Basic Books: New York, 2020).

Hribal, J. *Fear of the Animal Planet: The Hidden History of Animal Resistance*. (Petrolia CA: Counterpunch, 2011).

Milbank, J. Aquinas' Natural Law Isn't Only About Human Beings. *Church Life Journal* (2021).

McGary, H. *The Concept of Resistance: Black Resistance During Slavery*. Freedom, Equality, Social Change, Peden C, Sterba J, Mellen E. (New York: Edwin Mellin Press, 1989): 359-371.

Singer, P. *Animal Liberation*. (Harper, 1977).

Vasanthakumar A. Recent Debates on Victims' Duties to Resist Their Oppressors. *Philosophy Compass* (2020): 1-10.

Von Essen E, Allen M. Skill or Slaughter in "Fair Chase: What does Animal Resistance Tell us about Modern Sports Hunting?" *Between the Species* (2022): 149-167.

Von Essen, E, Allen M, Tickle L. Game of Drones: On the Moral Significance of Deception in Modern Sport Hunting. *Philosophical Journal of Conflict and Violence* (2020) 4:2: 137-157.

# Contents of Earlier Volumes

## Advances in Animal Science and Zoology. Volume 20


**Chapter 1** **Zebrafish Models of Schizophrenia: Gaining Insights into the Link from Genes to Brain to Behavior**
S. Y. Mok and P. Y. Tang

**Chapter 2** **Melatonin: Pleiotropic Action for the Regulation of Fish Reproduction in Aquaculture**
Joydeep Das, Akash Acharyya and Kazi Nurul Hasan

**Chapter 3** **Biotic Interaction Gradients between Native and Invasive Aquatic Gastropods**
Kiza K. Gates and Billie L. Kerans

**Chapter 4** **Competitive and Facilitative Mechanisms of a Biotic Interaction Gradient between Native and Invasive Aquatic Gastropods**
Kiza K. Gates and Billie L. Kerans

**Chapter 5** **The Role of Superoxide Dismutase in Reptiles under Toxicity Contexts**
L. M. Odetti, E. V. Paravani, M. F. Simoniello and G. L. Poletta

**Chapter 6** **Anasarca in English Bulldogs**
B. Valladares-Carranza, J. L. Zamora-Espinosa, M. G. Arriaga-Pérez, C. Ortega-Santana, Y. E. Felipe-Pérez, A. Zaragoza-Bastida, N. Rivero-Pérez, L. Delgadillo-Ruiz, R. Bañuelos-Valenzuela, P. I. Gallegos-Flores, V. Velázquez-Ordoñez and C. Bedolla-Cedeño

## Advances in Animal Science and Zoology. Volume 19

**Chapter 1** *Bemisia tabaci:* **Effects of Chemicals and Entomopathogens on Behaviour**
Ayub Khan

**Chapter 2** **Canine Lymphoma: A ReviewError! Bookmark not defined.**
Teresa Pinto and Felisbina Luisa Queiroga

**Chapter 3** **Diversity, Threats and Its Behavior on Freshwater Fishes**
Haryono and Rudhy Gustiano

**Chapter 4** **Comparative Study of Endocrine and Enzymatic Functions on Complex Behavior of Freshwater Fishes**
Jiban Kumar Behera, Manojit Bhattacharya, Akansha Mishra, Pabitra Mishra, Bhaskar Behera, Niladri Bhusan Kar and Bidhan Chandra Patra

**Chapter 5** **Contamination and Fish Behavior: The Conspicuous Case of Endocrine Disruptors (EDs)**
Sandra Isabel Moreno Abril

**Chapter 6** **Lower Urinary Tract Disease in Guinea Pigs (*Cavia porcellus*): A Review**
Salomé Azevedo and Felisbina Luisa Queiroga

**Chapter 7** **Parasitic Helminths of Introduced Marmosets That Cohabit the Area of Occurrence of the Golden Lion Tamarins in Brazil**
Gildeíde Aparecida Costa, Carlos Ramon Ruiz-Miranda and Clóvis de Paula Santos

## Advances in Animal Science and Zoology. Volume 18

**Chapter 1** **A Review of Camel Parasites Found in the UAE**
Rolf K. Schuster, Saritha Sivakumar, Jörg Kinne, Elisa Margarida Maio, Marina Ringo, Rekha Raghavan and Ulrich Wernery

**Chapter 2** **Antimicrobial Resistance in Piggery Farming**
Rajib Deb, Prassana Pal, Ajay Kumar, Souvik Paul, Seema Rani Pegu, B. C. Das and V. K. Gupta

**Chapter 3** **Behavioural Assays to Analyse the Muscle Mutants of *Drosophila melanogaster***
Ushashi Ain and Hena Firdaus

**Chapter 4** **The Adipokinetic Hormones of Cicadomorpha (Hemiptera, Auchenorrhyncha): First Steps for the Development of a "Green" Insecticide**
Gerd Gäde, Simone König and Heather G. Marco

**Chapter 5** **Aphid Major Agricultural Insect Pest, Ecology, Lifecycle, and Biological Control Approaches**
Khadija Javed, Humayun Javed and Dewen Qiu

**Chapter 6** **Foraging Behavior, Guild Structure, and Species Diversity in Australian Passerines**
Vladimír Remeš, Eva Remešová, Nicholas R. Friedman, Beata Matysioková and Lucia Rubáčová

**Chapter 7** **Mouth Structure and Nutrition Behaviour of Aphids**
Işıl Özdemir and Yunus Bayram

**Chapter 8** **When It's the Test That Fails: Why Gorillas May Not Match the Sample**
Jordyn Truax and Jennifer Vonk

## Advances in Animal Science and Zoology. Volume 17

**Chapter 1** **Breeding of Silkworm, *Bombyx mori*, in India**
Rajendra Mundkur

**Chapter 2** **Relationship between Mineral Licks and Terrestrial Mammals in the Tropical Rainforest of Malaysia: A Perspective Review**
Wing Shen Lim and Andy Russel Mojiol

**Chapter 3** **Prenatal Undernutrition in Farm Animals: Effects on Productivity and Behavior**
Panagiotis E. Simitzis and Maria A. Charismiadou

**Chapter 4** **Kin Preferences in Female Straw Colored Fruit Bats (*Eidolon helvum*)**
Jennifer Vonk

**Chapter 5** **The Pleiotropic Functions of Oncorhynchus mykiss Red Blood Cells**
Maria del Mar Ortega-Villaizan, Maria Elizabhet Salvador-Mira, Sara Puente-Marin, Ivan Nombela and Veronica Chico

**Chapter 6** **Amphibian Species Discovery Pattern in India: Past, Present and Future Trends**
K. P. Dinesh, C. Radhakrishnan and Nirmal U. Kulkarni

## Advances in Animal Science and Zoology. Volume 16

**Chapter 1** **Entomopathogenic Nematodes as Biological Control Agents in Benin: A Review**
Hugues Baimey, Régina Kotchofa, Lionel Zadji, Anique Godjo, Léonard Afouda, André Fanou, Régis Ahissou, Néhal Djaouga Mamadou and Wilfrieda Decraemer

**Chapter 2** **Silver Nanoparticles with Mosquito Control Potential: Optimal Synthesis and Biophysical Characterization**
Aarti Sharma, Monika Mishra, Vinay Singh Dagar and Sarita Kumar

**Chapter 3** **Functional Morphology of the Metathorax and Hind Wing of *Apis mellifera* (Hymenoptera: Apidae)**
Hans Klaus Pfau

**Chapter 4** **Advancements in Regenerative Engineering for Treating Urologic Dysfunction**
Bonnie G. Nolan and Arun K. Sharma

**Chapter 5** **Gas Production Technique as a Powerful Tool for the Evaluation of the Nutritional Quality of Feedstuffs in Ruminants' Production**
Manuel Murillo-Ortiz, Damián Reyes-Jáquez, Esperanza Herrera-Torres and Gerardo Pámanes-Carrasco

**Chapter 6** **Resurgence of Spodoptera Frugiperda (Lepidoptera: Noctuidae) in Maize Fields in Northern Benin: Towards the Development of a Control Method Integrating Entomopathogenic Nematodes**
Hugues Baimey, Régina Kotchofa, Lionel Zadji, André Fanou, Rufin Dossou-Agbede, Hamidate Euridice Tella and Anne-Marie Alia

# Index


## A

Altai Mountains, 56, 57, 58, 59, 60
analogical reasoning, 139, 140, 147, 149, 150, 151
analogy, 143, 144, 145, 187, 188, 191
Andaman and Nicobar Islands, 97, 98, 100, 111, 112
animal liberation, vii, 140, 148, 152, 187, 188, 193
animal resistance, vii, 152, 187, 188, 189, 191, 192, 193
antioxidant capacity, 35, 38, 39, 46, 98, 101, 107

## B

bighorn sheep, 78, 92, 93, 94, 95
biochemical profiles, 97, 99, 101, 104, 105
bovine respiratory syncytial virus (BRSV), 38, 39

## C

cAMP, v, vii, 117, 118, 125, 126, 127, 128, 129, 130, 131, 132
Carabidae, vi, vii, 153, 154, 155, 159, 161, 162, 163, 164, 165, 168, 169, 170, 171, 172, 173
cattle, v, vii, 33, 34, 35, 36, 37, 39, 40, 42, 43, 44, 45, 46, 47, 48, 50, 51, 52, 53, 94, 97, 98, 99, 103, 108, 109, 111, 112, 113, 114, 115, 116, 176, 180, 183, 184
Cerebrospinal Fluid Analysis (CSF), 6, 15, 23
chemotherapy, 2, 16, 20, 21, 22, 26, 27, 31, 113
choroid plexus tumor, vii, 1, 2, 3, 6, 8, 12, 13, 14, 15, 18, 21, 25, 26, 27, 28, 29, 31
cognitive ethology, 139, 140, 147
computed tomography (CT), 1, 10, 12, 21, 25, 119, 127
Continental Divide (CD) bighorn, 77, 79, 80, 81, 82, 83, 84, 85, 86, 87, 88, 89, 90, 91

## D

DAPA, 77, 78, 81, 85, 86, 88, 89, 90, 91, 95
diet quality, vii, 77, 78, 79, 81, 82, 85, 86, 87, 89, 90, 91, 94
diets, vii, 33, 34, 36, 39, 41, 42, 47, 48, 51, 52, 77, 78, 79, 81, 82, 83, 85, 86, 87, 88, 89, 90, 91, 94, 95, 185
Diets, 85, 87, 88
diversity, 71, 95, 153, 154, 155, 159, 160, 161, 169, 170, 172, 173, 196, 197
dog, 1, 2, 25, 26, 27, 29, 74

## E

empathy, 152, 187, 188, 191, 192

## F

fecal nitrogen, 78, 81, 86, 93, 94

## G

geochemical characteristics, v, 55, 56, 69
geoecology, 56, 70
geological formations, v, 55, 56, 60, 61, 63, 65
glioma, 2, 7, 8, 11, 13, 18, 21, 25, 27
goat diseases, 176

Gorny Altai, v, 55, 57, 58, 59, 60, 69, 71, 72, 73

## H

haemonchosis, 176, 182, 183, 184
*Haemonchus contortus* (*H. contortus*), vii, 175, 176, 177, 179, 180, 181, 182, 183
hormone profiles, 101
humpsore, v, vii, 97, 98, 100, 102, 103, 104, 105, 106, 107, 108, 109, 110, 111, 112, 113

## I

immune function, 34, 35, 37, 38, 39, 46, 51
index, vi, 14, 23, 28, 58, 74, 77, 79, 81, 82, 87, 88, 90, 94, 153, 154, 159, 160, 168, 169, 201

## L

literature review, 2

## M

macrocyclic lactone, v, vii, 97, 98, 99, 108
mammals, 43, 55, 56, 57, 60, 61, 63, 64, 66, 67, 68, 69, 70, 71, 72, 73, 74, 75, 77, 92, 94, 184, 198
maturation-inducing hormone (MIH), 117, 118, 119, 120, 121, 123, 124, 125, 127, 128, 133
membrane progesterone receptor (mPR), 117, 118, 120, 121, 123, 124, 125, 126, 128, 130, 131, 132, 136, 137
meningioma, 2, 7, 8, 12, 17, 18, 19, 20, 22, 24, 26, 27, 28, 29, 30
morphological identification, 176
Mummy Range (MR) bighorn, 78, 82, 85, 87, 88, 89, 90, 91
myogenesis, 34, 35, 38, 43, 52

## N

natural habitat, 153, 155, 156, 161, 162, 163, 164, 166, 168, 169, 171
natural law, vii, 139, 140, 141, 142, 143, 144, 145, 146, 147, 148, 149, 150, 151, 152, 190, 192

## O

oocyte maturation, v, vii, 117, 118, 119, 121, 123, 124, 125, 126, 127, 128, 129, 130, 131, 132, 136
oxidative stress, 39, 46, 98, 101, 102, 109, 111, 112

## P

piperazine derivative, 97, 98, 99
population performance, 78, 79, 92
preadipocyte formation, 34, 35, 38, 43, 53
progesterone, 22, 23, 24, 118, 119, 121, 123, 125, 126, 127, 128, 129, 130, 131, 132
progestin and AdipoQ receptor (PAQR), 118, 124, 132

## R

radiotherapy, 2, 3, 16, 17, 19, 24
reproductive function, 41
requirement, 34, 36, 39, 46, 50
retinol, 33, 34, 35, 36, 37, 38, 39, 41, 42, 43, 46, 47, 48, 49, 51, 52, 53
Rocky Mountain bighorn sheep, v, vii, 77, 78, 92, 95
Rocky Mountain National Park (RMNP), vii, 77, 79, 80, 81, 83, 86, 89, 90, 91, 92, 93, 94

## S

Scarabaeidae, vi, vii, 153, 154, 155, 159, 161, 166, 167, 168, 171, 173
semi natural habitat, 153

steroid hormones, 117, 118, 120, 121
suffering, 110, 149, 187, 188, 189, 191, 192
supplementation, 30, 33, 34, 35, 38, 39, 40, 41, 42, 43, 45, 46, 47, 48, 49, 111, 115, 177
surgery, 2, 12, 16, 17, 19, 24, 25, 26, 27, 51, 116

## T

therapeutic agents, 99, 113
total liberation, vi, 139, 140, 143, 144, 145, 147, 148, 149, 150, 151, 152

## U

urbanization area, 169
urbanization effect, 154, 171, 172

## V

vitamin A, v, vii, 33, 34, 35, 36, 37, 38, 39, 40, 41, 42, 43, 44, 45, 46, 47, 48, 49, 50, 51, 52, 53
vulvar flap, 175, 176, 177, 178, 179, 180, 182

## W

World Health Organization (WHO), 12, 13, 28, 114

## β

β-carotene, 36, 40, 43, 49, 50, 51, 52